Covalent Organic Frameworks

Edited by Yanan Gao and Fei Lu

Published in London, United Kingdom

Covalent Organic Frameworks
http://dx.doi.org/10.5772/intechopen.102264
Edited by Yanan Gao and Fei Lu

Contributors
Yanan Gao, Fei Lu, Tao Zhang, Yuxiang Zhao, Guiyang Zhang, Shuai Bi, Yingjie Zhao, Hui Liu

First published in London, United Kingdom, 2023 by IntechOpen
IntechOpen is the global imprint of INTECHOPEN LIMITED, registered in England and Wales, registration number: 11086078, 5 Princes Gate Court, London, SW7 2QJ, United Kingdom

British Library Cataloguing-in-Publication Data
A catalogue record for this book is available from the British Library

Additional hard and PDF copies can be obtained from orders@intechopen.com

Covalent Organic Frameworks
Edited by Yanan Gao and Fei Lu
p. cm.
Print ISBN 978-1-80356-959-8
Online ISBN 978-1-80356-960-4
eBook (PDF) ISBN 978-1-80356-961-1

Meet the editors

Yanan Gao is the vice director of the Key Laboratory of Advanced Materials of Tropical Island Resources, Ministry of Education, at Hainan University, China. He received his bachelor's degree in 1999 and Ph.D. in 2005 from Shandong University. He continued his research at the Max Planck Institute for Dynamics of Complex Technical Systems (2006-2008), University of Akron (2008-2009) and the University of York (2009-2011) as a postdoctoral research fellow. In 2011 he joined the Dalian Institute of Chemical Physics, Chinese Academy of Sciences as a principal investigator in the Ionic Liquids Lab and the leader of the Porous Organic Framework Material Group. His current research interests focus on the design, synthesis and application of covalent organic frameworks (COFs).

Fei Lu is a professor at the School of Chemical Engineering and Technology at Hainan University. She received her bachelor's degree in material chemistry from Qufu Normal University in 2010 and a Ph.D. in colloid and interface chemistry from Shandong University in 2015. From 2015 to 2019, she was a postdoctoral research fellow at Shandong University and the University of Waterloo. She was promoted to full professor at Shandong Normal University in 2019. Her current research interests are in porous materials for energy storage.

Contents

Preface

Covalent organic frameworks (COFs) provide a molecular platform for the production of predictable ordered and extended 2D or 3D structures with topologically well-defined channels and discrete micropores and/or mesopores, greatly enhancing the possibilities of transforming organic materials into unique properties and functions. The past 17 years have seen tremendous achievements in the design, synthesis and functionalization of COFs to show their promising potential as a class of fascinating organic materials. By way of illustration, this book first summarizes recent advances in the interfacial synthesis of 2D COF films via gas/solid, liquid/liquid, liquid/solid and gas/liquid interface. The COF films obtained exhibit better orientation and fewer defects, endowing them with considerable potential for further applications. Next, recent progress in COFs for photocatalysis in water splitting, CO_2 reduction, organic transformation and environmental pollutant degradation is discussed. The latest achievements of COFs as solid ion conductors in energy devices are summarized, the controlled size, topology and interface properties of their ordered pores offering ideal pathways for long-term ion conduction. Finally, the possibility of COFs as emerging platforms in cancer therapy is discussed. Encouragingly, successful cases including chemotherapy, photodynamic therapy, photothermal therapy and immunotherapy have been realized. We believe this book will give readers a broad overview of the design criteria and practical methodologies of COFs in various application fields.

Yanan Gao and Fei Lu
Hainan University,
Haikou, China

Chapter 1

Interfacial Synthesis of 2D COF Thin Films

Tao Zhang and Yuxiang Zhao

Abstract

Two-dimensional covalent organic frameworks (2D COFs) are emerging crystalline 2D organic material comprising planar and covalent networks with long-ranging structural order. Benefiting from their intrinsic porosity, crystallinity, and electrical properties, 2D COFs have displayed great potential for separation, energy conversion, and electronic fields. For the most of these applications, large-area and highly-ordered 2D COFs thin films are required. As such, considerable efforts have been devoted to exploring the fabrication of 2D COF thin films with controllable architectures and properties. In this chapter, we aim to provide the recent advances in the fabrication of 2D COF thin films and highlight the advantages and limitations of different methods focusing on chemical bonding, morphology, and crystal structure.

Keywords: interfacial synthesis, 2D material, COF, crystal, thin film

1. Introduction

In 2005, the first covalently bonded crystalline porous polymer was successfully synthesized and named covalent organic frameworks (COFs). In subsequent developments, COFs linked by B−O, C=N, C=C, and other bonds have been reported. The regular network structure of COFs can be fully characterized with the help of existing instruments, which is very beneficial to study the relationship between the performance and structure. COFs can also be divided into two-dimensional (2D) COFs and three-dimensional (3D) COFs according to the specific structure. 2D COFs are emerging crystalline 2D organic material comprising planar and covalent networks with long-ranging structural order [1, 2]. In recent years, 2D COFs have been rapidly developed due to their ease of synthesis and definite structure. Benefiting from their intrinsic porosity, crystallinity, and electrical properties, 2D COFs have displayed great potential for separation [3, 4], energy conversion [5–7], and electronic fields [8]. The preparation of COF materials as thin films is advantageous for most applications. Large-area and highly ordered 2D COFs thin films are widely studied [9–12]. COFs are mostly connected by reversible covalent bonds. If the reaction conditions can be adjusted to make the structure of COFs in a dynamic self-repair process, highly ordered films can be obtained. Clever use of the interface can also give the film a good substrate for growth, and the COFs can be spread out along the interface to produce a smooth film. Common interfaces include gas/solid interface, liquid/liquid interface, liquid/solid interface, and gas/liquid interface. Hence, in this critical review, we aim to provide the recent advances in the fabrication of 2D COF thin films and highlight the advantages and limitations of different methods focusing on chemical bonding, morphology, and crystal structure.

2. Preparation methods of COF film

At present, the common preparation methods include top-down method and bottom-up method. The bottom-up method is mainly through the interface reaction, so that the formation and rupture of chemical bonds of small organic molecules occur at the interface, and the final product is spread along the interface. The interfaces include gas/solid interface, liquid/liquid interface, liquid/solid interface, and gas/liquid interface. 2D COF films obtained by this method have better orientation and fewer defects, and the relationship between properties and structure can be better studied in the application. But there are some problems such as low yield and slow reaction through interface reaction. Reaction time and efficient utilization of small molecular monomers are also important issues to be considered in the process of synthesizing materials. The top-down preparation method is to peel the powder COFs material into nanosheets by chemical treatment or physical method and then process the nanosheets into large-size films by vacuum extraction and filtration. The prepared materials via this method have many defects, weak orientation, and other problems. How to control the thickness and structure of materials to obtain 2D films with orderly atomic structure is the main challenge at present (**Figure 1**).

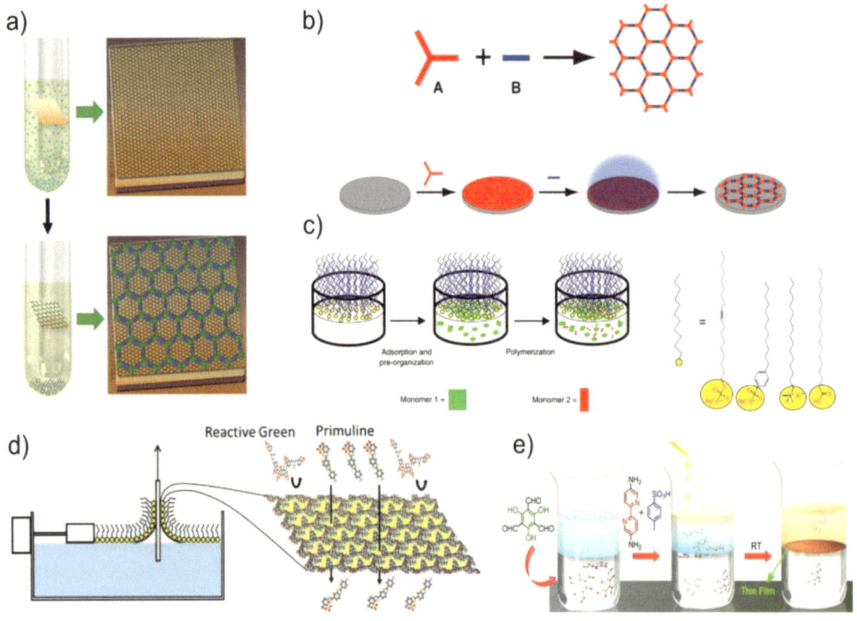

Figure 1.
Schematic illustration of interface synthesis of COF films. (a) Solid/liquid interface; (b) air/solid interface; (c) air/liquid interface; (d) Langmuir-Blodgett (LB) method; and (e) liquid/liquid interface.

2.1 Bottom-up strategy

2.1.1 Solid/liquid interface

COFs were first reported by Yaghi et al. in 2005. COF-1 and COF-5 connected by B-O bond were respectively prepared in Pyrex tubes by solvothermal method [1]. The reversible B-O bond gives the material excellent repairability, resulting in a highly regular pore structure and a high specific surface area. Solvothermal synthesis is still the main strategy for preparing COFs powder [13–15].

The preparation of COF film by gas/solid interface is mainly to add solid substrate in the solvent and let COFs grow into film on the surface of the substrate in situ. In 2011, Dichtel et al. prepared COF-5 films from the polymerization of two monomers, 2,3,6,7,10,11-hexahydroxytriphenyl (HHTP) and 1, 4-phenyldiboric acid (PBBA), using a single layer of graphene as a substrate [16] (**Figure 2a**). In this process, the concentration of monomer needs to be regulated. When the concentration of monomer is high, the system will nucleate and generate powder. The characterization of the COF-5 film by grazing incidence X-ray diffraction (GIXRD) and scanning electron microscope (SEM) showed that the films were parallel to the graphene surface and have good orientation (**Figure 2b–d**). In this work, COF powder was processed into film for the first time, which is of great significance for the application of COF as organic electronic devices. When COF is processed into a film, it can better characterize its optoelectronic properties. Inspired by this work, different research groups began to choose different substrate materials for the preparation of thin films [17–19]. Base materials include Indium Tin Oxides (ITO) and polyethersulfone (PES) [20, 21], etc. The synthetic conditions of BDT-ETTA COF and the transmission electron microscopy (TEM) images of the powder are shown in **Figure 3a** and **b**, respectively. SEM images and GID patterns analysis of the cross section found that the COF film was spread evenly on the ITO substrate (**Figure 3c–f**). The connecting units of COFs include

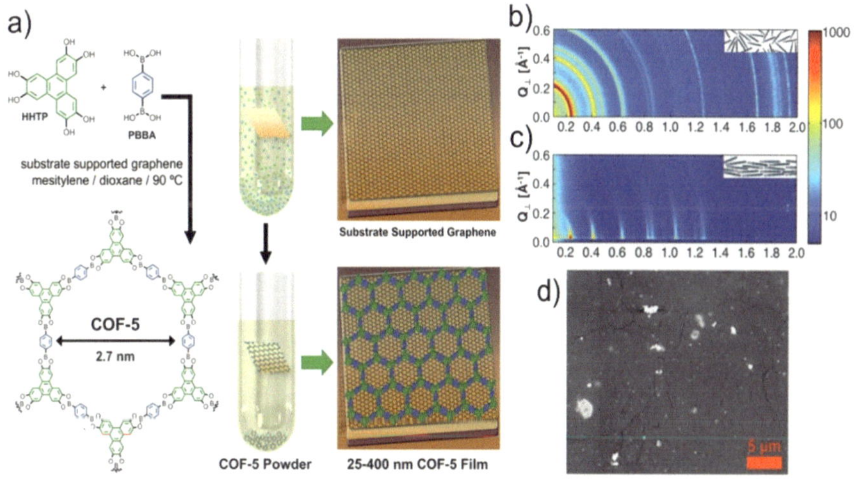

Figure 2.
(a) Schematic illustration of chemical synthesis at solid/liquid interface using a single layer of graphene as a substrate; (b) X-ray scattering data obtained from COF-5 powder; (c) GID data from a COF-5 film on SLG/Cu; (d) top-view SEM image of the COF-5 thin film.

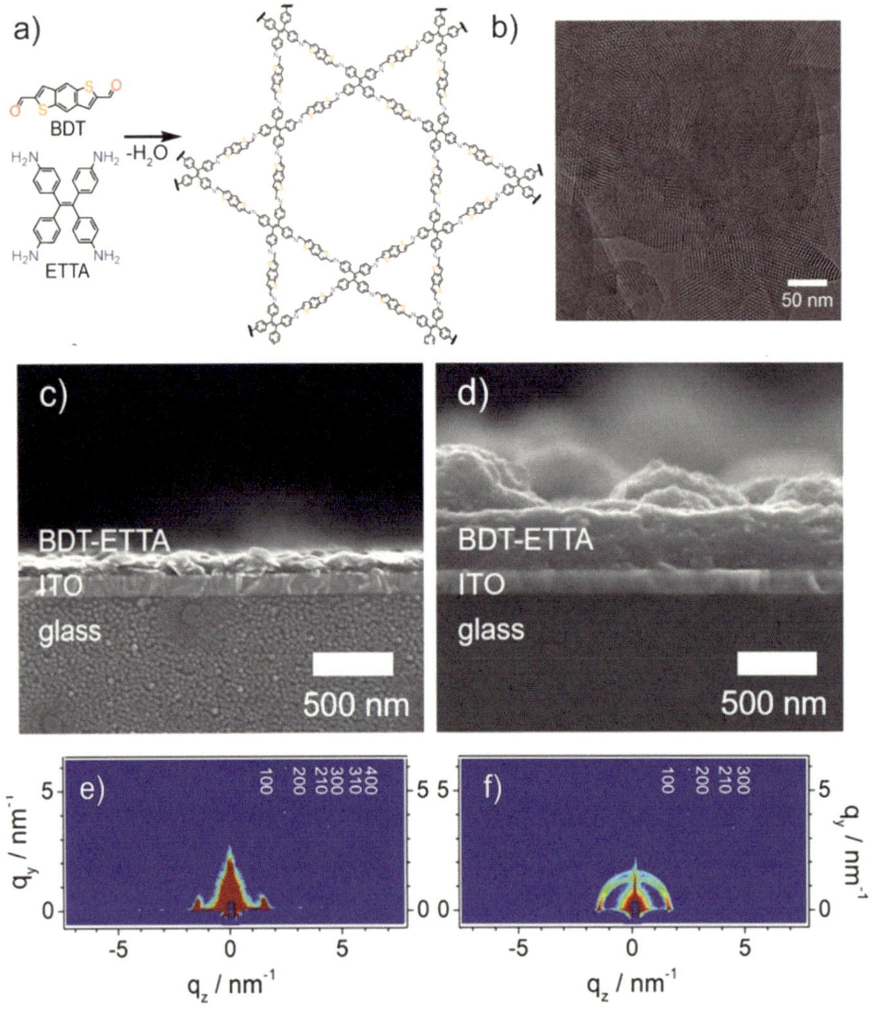

Figure 3.
Schematic illustration of chemical synthesis at solid/liquid interface using the indium tin oxides (ITO) as a substrate; (a) synthesis process of BDT-ETTA COF; (b) TEM image of BDT-ETTA COF powder; (c,d) SEM images (cross-section) of COF film; (e, f) GID patterns of COF film.

B–O bond and C=N bond. These thin films can be well used for water splitting and molecular separation. Solid/liquid interface synthesis is considered to be an effective method for vertically growing high crystallinity films. Since the film grows on the substrate surface, it is easy to collect and characterize. However, there is another problem with this method, that is, the adhesion force between the film and the substrate is strong, and it is challenging to transfer it from the substrate. Since the film cannot be transferred between different substrates, it is only possible to select the appropriate substrate before each experiment [22].

2.1.2 Liquid/liquid interface

Water and various organic solvents are incompatible, resulting in a phase interface at the junction of the two phases. If COFs can be spread out during growth, 2D COF films with excellent orientation can be obtained at the interface with the continuous self-repair of dynamic chemical bonds. The films obtained by this method can be easily transferred to different substrate materials, which is very beneficial in application [23, 24]. In 2016, Xinliang Feng et al. designed and synthesized wafer-sized 2D imine COFs with high mechanical stiffness. Porphyrin amino monomer and 2,5-dihydroxyterephthalaldehyde monomer are dissolved in trichloromethane and water, respectively [25] (**Figure 4a**). The amine group and aldehyde group contact at the interface to form imine bond, leading to 2D polymerization. The authors simulated the structure of COF film and characterized the surface topography of the material by atomic force microscopy (AFM) and TEM, in which a highly ordered arrangement of structures can be seen (**Figure 4b–d**). Photographic image of 2DP on 4-inch 300 nm SiO_2/Si wafer shows that the entire thin film is macroscopically flat (**Figure 4e**). Liquid/liquid interface synthesis is another method that can efficiently synthesize high-quality films. At present, the main problem is to prepare COF films with higher crystallinity by adjusting the amount of catalyst and the choice of solvent system. On the basis of this work, more efficient catalysts of Schiff base reaction catalysts have also been developed. Dichtel et al. prepared high crystallinity COF films efficiently using $Sc(OTf)_3$ as a catalyst [26, 27]. The catalyst is dissolved in the aqueous phase, and the monomers are dispersed in the organic phase. Amine and aldehyde monomers can be polymerized into films at the phase interface under the catalysis of $Sc(OTf)_3$ (**Figure 4f–g**). Bo Wang et al. also used this catalyst to covalently connect 2, 5-dihydroxy-1, 4-phthalate formaldehyde (DOBDA) and 1,3,

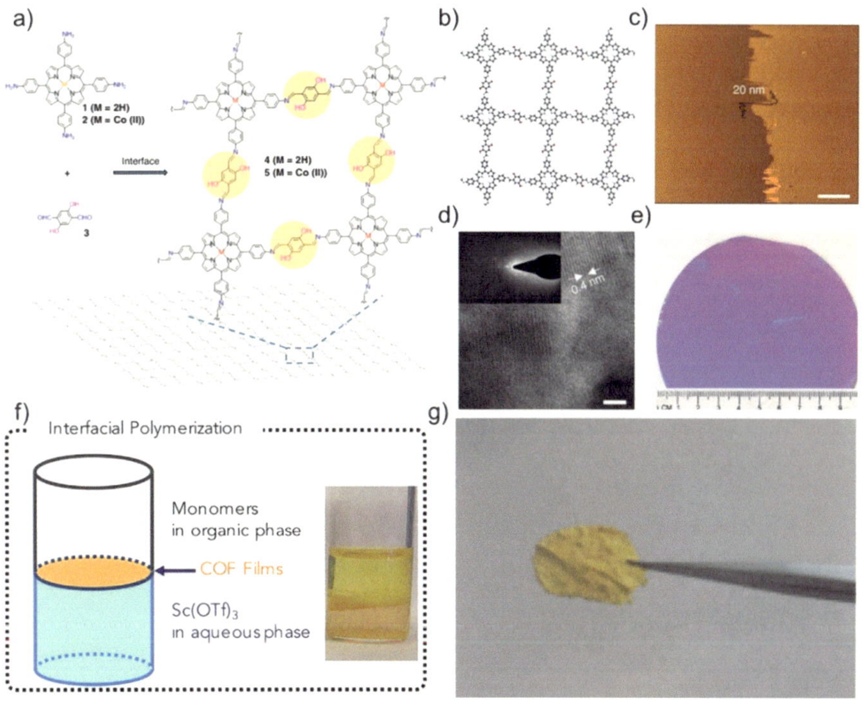

Figure 4.
Schematic illustration of chemical synthesis at liquid/liquid interface. (a) Synthesis of a 2DP through Schiff-base condensation reaction; (b) molecular structure of the 2DP by DFTB calculation. (c) AFM, and (d) TEM images of 2DP; (e) photographic image of 2DP on 4-inch 300 nm SiO_2/Si wafer; (f) schematic explanation and photograph of TAPB-PDA COF; (g) photo of the TAPB-PDA COF film.

5-tri (4-aminophenyl)-benzene (TAPB) to prepare compact COF films with different thickness of 300 nm ~ 500 nm [28]. Besides, by removing C=N bonds and using the defects in COF films, a vertical channel with hydrophilic gradient was fabricated.

The reaction time of interfacial synthesis is another problem that needs to be considered. In order to obtain high crystalline films, the rate of reaction is usually controlled to slow down the polymerization kinetics. In 2019, wafer-scale synthesis of monolayer 2D porphyrin polymers was reported by Park Research Group (**Figure 5a**). The authors report that by growing the film at the pentane/water interface, a 2D porphyrin polymer film with sheet-level uniformity was synthesized at the limit of the thickness of a single layer, by growing the film at the pentane/water interface [29]. Films of different structures have different absorption spectra and colors (**Figure 5b–d**). The superposition of films of different colors also reveals

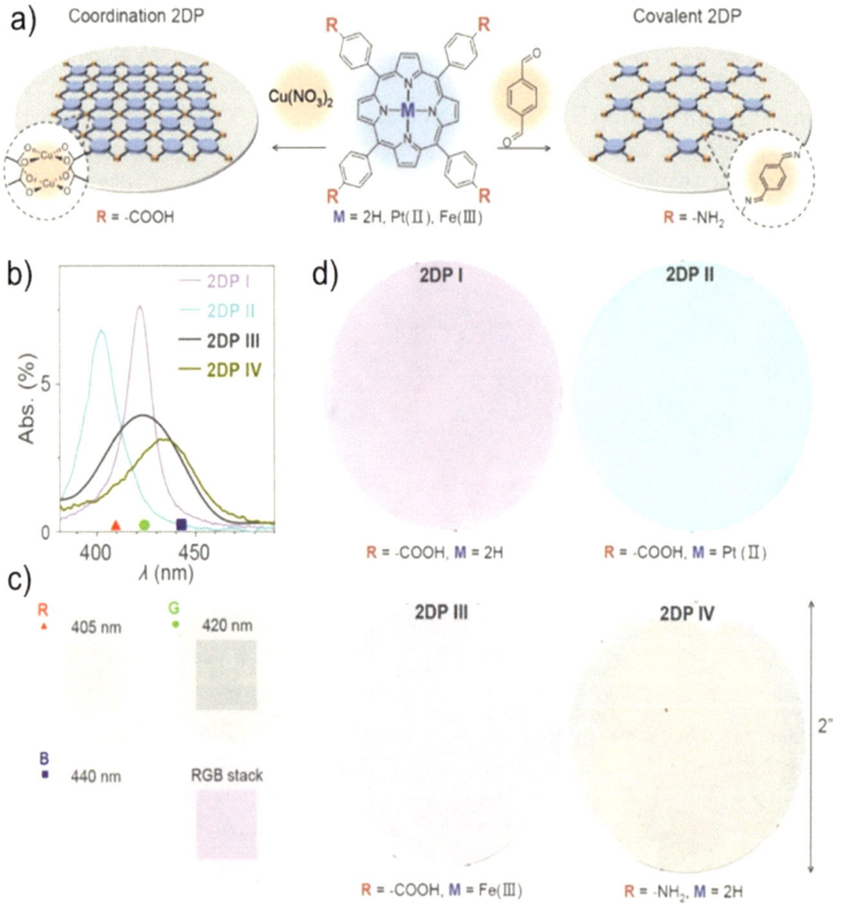

Figure 5.
(a) Schematic of monolayer 2DPs and corresponding chemical structures of the molecular precursors; (b) absorption spectra of monolayer 2DPs on fused silica substrates; (c) hyperspectral transmission images and resulting false color images of 1 inch-square 2DP I on a 2-inch fused silica substrate. Transmission images taken at the wavelength of 405 nm, 420 nm, and 440 nm are assigned as red, green, and blue channel, respectively, to generate the false color image. (d) False color images of monolayer 2DPs covering entire 2-inch fused silica wafers.

different optical properties. The corresponding color and monomer are very close, which provides experience for future film designs of different colors.

In 2017, Banerjee et al. selected p-toluenesulfonic acid (PTSA) to form a self-supporting COF film at the interface between water and methylene chloride [30] (**Figure 6a**). The hydrogen bond network formed by PTSA can slow down the diffusion rate of monomer and increase the quality of the COF film. The content of catalyst and concentration of monomer have great influence on the thickness of thin films. SEM image and AFM image of Tp–Bpy COF thin film prove that the surface of the film is very smooth and the structural orientation is high (**Figure 6b–c**). Zhongyi Jiang et al. prepared ionic covalent organic framework membranes (iCOFMs) with ultrahigh ion exchange capacity via the double-activation interfacial polymerization strategy (**Figure 6d**). Brønsted acid and Brønsted base activate aldehyde monomers and amine monomers with sulfonic groups in aqueous and organic phases, respectively [31]. After the double activation of acid and base, the monomer can react quickly at the interface and form iCOFMs with high crystallinity. At present, the liquid/liquid interface synthesis is developing rapidly. The films synthesized by this method are widely used in the fields of separation and purification and seawater desalination. At the same time, this method has also been continuously improved in the process of development, which can obtain higher quality films in a shorter time.

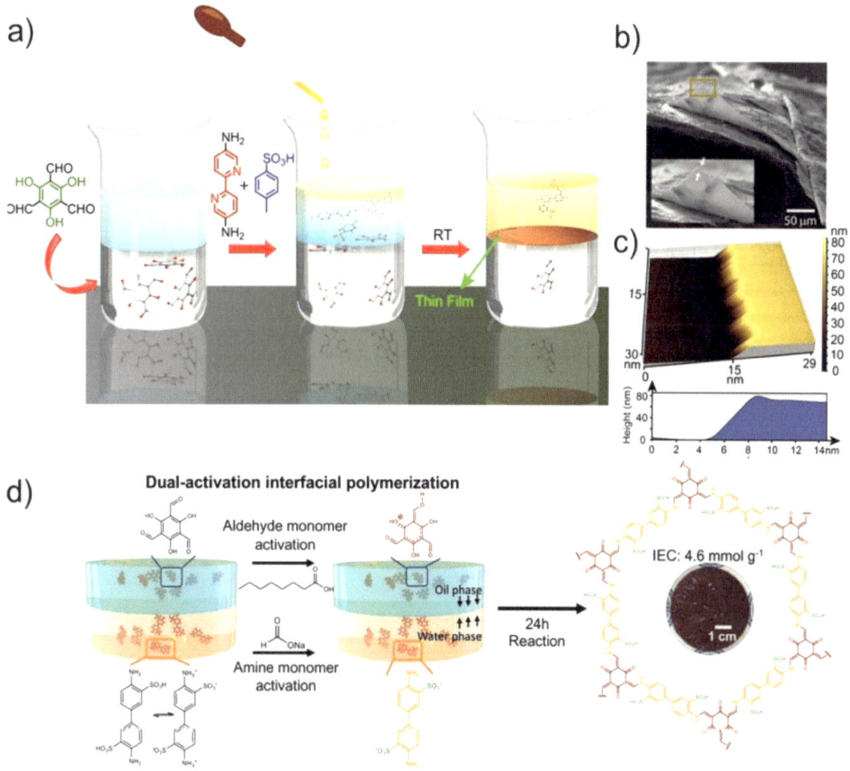

Figure 6.
(a) Schematic illustration of the interfacial synthesis of Tp–Bpy COF thin film; (b) SEM images of Tp–Bpy COF thin film; (c) AFM image of Tp–Bpy COF thin film; (d) schematic illustration of the TpBD-$(SO_3H)_2$ iCOFMs fabrication process.

2.1.3 Air/liquid interface

Zhenan Bao et al. first reported the preparation of polyTB film via DMF/air interface reaction in 2015 [32]. The two types of monomers are 4,8-Bis(octyloxy) thieno[2,3-f][1]benzothiophene-2,6-dicarbaldehyde (BDTA) and Tris(4-Aminophenyl)amine (TAPA). Since the COFs powder is easily formed in the direct polymerization process, the surface of the film obtained by this method is very rough. To solve this problem, the team continued to grow the film using the solution from the first reaction. By controlling the reaction conditions, the polyTB film with different thickness was obtained. The average surface roughness of the material is only 0.2 nm. Lai et al. loaded ultra-thin TFP-DHF 2D COF film (2.9 nm) on porous substrate via the Langmuir-Blodgett (LB) method. This COF was formed through the reaction between 1,3,5-triformylphloroglucinol (TFP) and 9,9-dihexylfluorene-2,7-diamine (DHF).

In 2019, Xinliang Feng et al. reported surfactant-monolayer-assisted interfacial synthesis (SMAIS) as a general method to prepare 2D polymer films with high crystallinity (**Figure 7a**). Sodium oleyl sulfate (SOS) was used as a surfactant to induce aniline to align and polymerize on the surface of aqueous solution and obtain fully conjugated

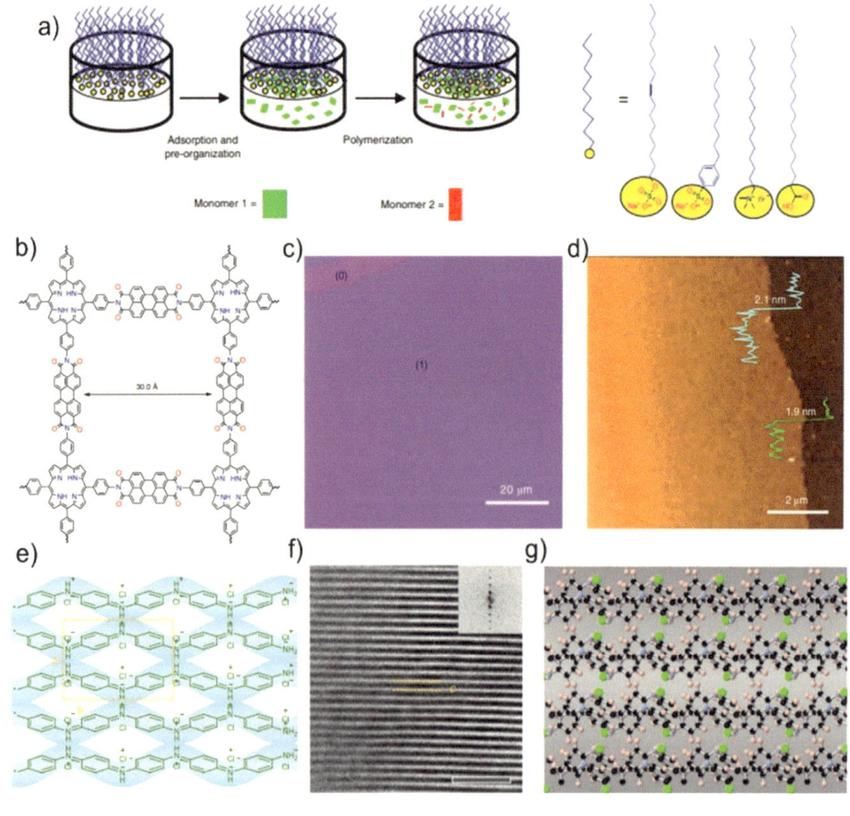

Figure 7.
Schematic illustration of chemical synthesis at air/liquid interface. (a) the synthetic procedure for the 2D polymers; (b) molecular structure of the 2DPI synthesized in the article; (c) optical microscope image of 2DPI film; (d) AFM image of the 2DPI film; (e) molecular structure of the q2D polyaniline synthesized in the article; (f) AC-HRTEM image of q2D polyaniline perpendicular to [001] axis; (g) simulated atomic structure of the q2D polyaniline.

2D polyaniline films with lateral size ~50 cm [2] and tunable thickness (2.6–30 nm) (**Figure 7e–g**) [33]. In another work, the researchers prepared polyimide COF film by the reaction of 4,4′,4′′,4′′′-(porphyrin-5,10,15,20-tetrayl)tetraaniline and disochromeno (**Figure 6b**) [34]. The polyimide COF film was characterized by X-ray scattering and AFM (**Figure 7c** and **d**). It is found that the thickness is about 2 nm and the average domain size is about 3.5 μm [2]. Later, on the basis of the above two works, Xinliang Feng et al. also synthesized a series of COF films connected by C=N and B−O bonds [35, 36]. Bien Tan et al. reported an aliphatic amine-assisted interfacial polymerization method to obtain independent covalent triazine frameworks (CTFs) films [37]. The structure was investigated by grazing incidence wide-angle X-ray scattering (GIWAXS) and small-angle X-ray scattering (SAXS). The lateral size of the film was up to 250 cm [2], and average thickness can be tuned from 30 to 500 nm. The unique conjugated structure of the material endows it with excellent photocatalytic performance for hydrogen evolution reaction. Zhikun Zheng et al. synthesized a series of imine COF films assisted by charged polymers [38]. The morphology and diffusion of preorganized monomers in charged polymers on water surface are very important for the formation of organic two-dimensional crystals. The monomers were 5,10,15,20-Tetrakis (4-aminophenyl)-21H,23H-porphyrin (TAPP) and four aldehyde monomers with different structures. Air/liquid interfacial synthesis is the most common method for preparing 2D COF films. The film is easy to operate and has good crystallinity and orientation. The air/liquid interface is very helpful for the preparation of single-layer or few-layer COF films. First, the monomers should be induced to arrange at the interface, and then the crystallinity of the film can be better guaranteed in the process of polymerization. So how to arrange the monomers on the surface is the main consideration in this method. The transfer and characterization of monolayers are also not easy. Generally, advanced electron microscope is needed to observe its morphology.

2.1.4 Air/solid interface

In 2008, Porte et al. prepared monolayer surface covalent organic frameworks (SCOFs) on the surface of Ag (111) via chemical vapor deposition (CVD) [39]. The polymerization process is achieved by the reaction of 1, 4-Benzenediboronic acid (BDBA) and boronic acid. The experiment was performed under ultrahigh vacuum (UHV) conditions. Two monomers were sublimated from two heated molybdenum crucible evaporators to the clean Ag (111) surface to obtain a molecular array.

Dong Wang et al. reported the formation of highly ordered 2D COF film via dehydration reaction of boronic acid (**Figure 8a**). The molecular layers were imaged at room temperature using scanning tunneling microscopy (STM) (**Figure 8b–c**). In 2013, Dong Wang et al. reported another method for the synthesis of imine COF films at the air/solid interface [40]. The solution of the two monomers was first coated on the substrate and then sealed in a reactor with copper sulfate pentahydrate as a thermodynamic regulator. By heating the reactor to a specified temperature to control the evaporation of aldehyde monomers, the aldehyde monomers condense on the surface covered by amine monomers and polymerize with them, high-quality monolayer imine COF films can be prepared at the air/solid interface [41]. Recently, Yunqi Liu et al. reported the preparation of large area imine PyTTA-TPA COF film with controllable thickness by gas-phase induced conversion in a CVD system (**Figure 8d**) [42]. The assembly process is achieved by reversible reaction between 4,4′,4″,4″′-(1,3,6,8-Tetrakis(4-aminophenyl)pyrene (PyTTA) film and terephthalaldehyde (TPA) vapor. Driven by π-π superposition and catalyzed by acid vapors,

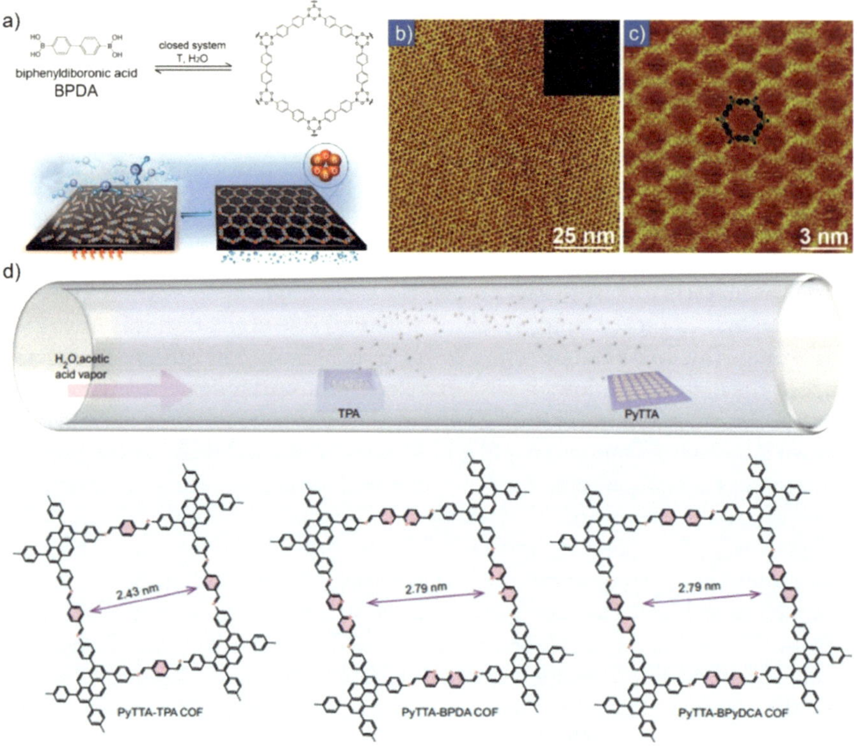

Figure 8.
Schematic illustration of chemical synthesis at air/solid interface. (a) the synthesis route to SCOF-1; (b) STM image of SCOF-1 on HOPG formed after dehydration of BPDA precursors at 150°C; (c) a high-resolution STM image showing the hexagonal structure of SCOF-1; (d) schematic representation for the growth of imine-linked 2D COF films on SiO_2/Si substrates.

a uniform organic frame film was formed on a growing substrate. The COF films obtained by air/solid interface synthesis have adjustable thickness and highly ordered structure, which is an effective method to grow high-quality thin films [43–45]. However, this method requires high temperature and vacuum environment and has high requirements for equipment.

2.1.5 Vapor-assisted conversion

Some organic solvents with low boiling point can evaporate at room temperature. Combined with this feature, Bein et al. reported the strategy of steam-assisted conversion at room temperature in 2014 [46]. They succeeded in making three thin films named BDT-COF, COF-5, and pyrene-COF (**Figure 9a**). The monomers are first mixed into a solution of acetone/ethanol, which is then dripped onto the matrix. Finally, the material is transformed into crystalline porous COF films in a vapor atmosphere of homotrimethylbenzene/dioxane. The presence of vapor plays an important role in the growth of thin films.

This method can accurately prepare COF films from 100 nm to a few microns thick. SEM characterization showed that the structure of the film was composed of small particles stacked irregularly, and there were submicron intervals between the particles (**Figure 9b** and **c**).

2.1.6 Continuous flow condition synthesis

Because COFs are easy to form powder particles under thermodynamic conditions, the films synthesized by liquid/solid interface method may have the problem of high surface roughness. To remedy this problem, Dichtel et al. converted the solution from a static state to a flowing state [47]. The flow rate affects the surface morphology, thickness, and crystallization degree of films. High-quality films with different thickness can be prepared by adjusting this condition (**Figure 10**). A quartz crystal microbalance can be used as the base of the flow tank to monitor the quality of film deposition at any time.

2.2 Other promising preparation methods of COF films

COFs powders are formed by layer upon layer of planes through π stacking and the interlayer forces are weak relative to covalent bonds. Similar to graphene, a large

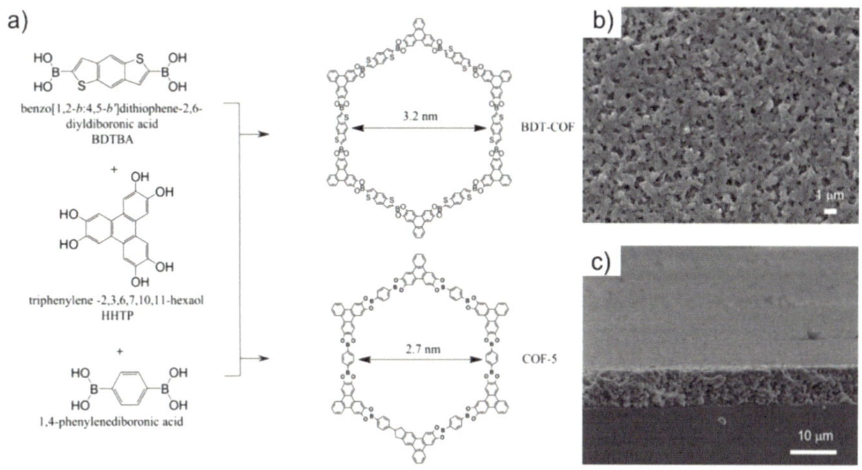

Figure 9.
(a) Schematic representation of BDT-COF and COF-5; (b) top view SEM micrograph of BDT-COF film synthesized by room temperature vapor-assisted conversion, representing the surface morphology; (c) cross-sectional SEM micrograph shows a uniform film thickness.

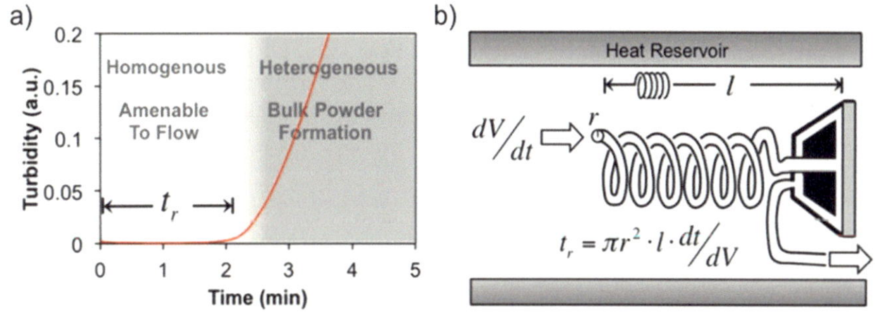

Figure 10.
(a) Turbidity as a function of reaction time during the formation of COF from homogeneous conditions provides an induction period amenable to a flow cell configuration. (b) Schematic of flow setup designed with variable induction period.

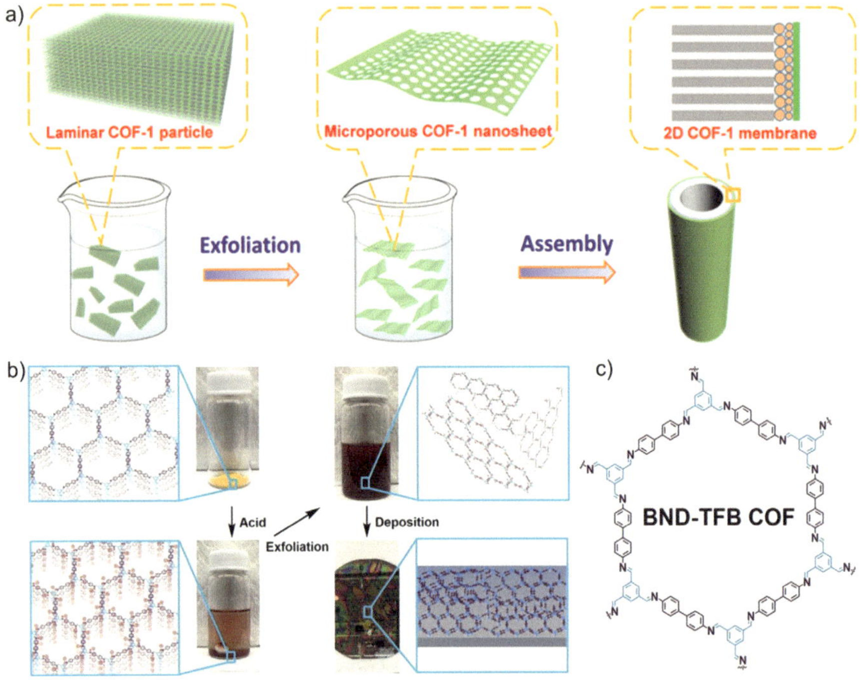

Figure 11.
(a) Schematic illustration of the preparation of a COF-1 membrane via the assembly of exfoliated COF-1 nanosheets; (b) overview of acid exfoliation and film casting procedures; (c) pore structure of the BND-TFB COF.

area of smooth COF film can be obtained if the powdered COFs can be stripped into a single layer of nanosheets, which can then be self-assembled into films [48]. The common preparation methods of nanosheets include physical exfoliation and chemical exfoliation [49, 50].

The physical method is to disperse COFs powder in the solvent and then form nanosheets assisted by ultrasound. Chemical methods require the addition of a chemical agent to the solvent to promote lamellar abscission. Compared with COFs powder, COFs nanosheets show more advantages in photoelectric applications. In addition, the processing of nanosheets into films is also a key point in practical applications. In 2017, Tsuru et al. applied the obtained COF-1 nanosheet solution drops on an α-Al_2O_3 macroporous support with a SiO_2-ZrO_2 intermediate layer and obtained uniform and smooth COF films after several drops (**Figure 11a**) [51]. Dichtel et al. protonated imine bond using trifluoroacetic acid to promote the stripping of COFs powder into nanosheets dispersed in a solvent (**Figure 11b** and **c**) [52]. COF films with thickness ranging from 50 nm to 20 μm can be prepared by deposition on any substrate.

3. Conclusion

The above is a review of common methods of film synthesis, including interfacial synthesis and some other new and promising preparation methods. By replacing different organic monomers, COF films with different functions can be designed in a targeted manner. Combined with the final different applications, the appropriate preparation method can be selected. At present, the solid/liquid interface synthesis method is mature and suitable for most COFs. However, if organic solvents are used for solvothermal reaction at high temperature, only inorganic materials can be selected as the substrate. COF films grown on solid surfaces cannot be transferred to other substrates. The air/solid interface synthesis also has the problem that the grown film is difficult to transfer. For liquid/liquid interface synthesis, organic solvents and aqueous solutions will form an obvious contact surface, which is an ideal substrate for the efficient growth of films at the interface. Liquid/liquid interface synthesis is also the most promising method to grow single crystal thin films. This method usually requires dissolving aldehydes and amines in two solvents, respectively, and reacting at the interface to form a film. The growth of COF thin films depends on the diffusion of monomers in solvents. The process of film formation at the interface will affect the diffusion of molecules, and the subsequent polymerization reaction will also be limited. Therefore, the surface of the grown film will be rough. If the two monomers are dissolved in the same organic solvent, and then the catalyst is dissolved in water, the monomers are catalyzed at the interface, and the polymerization reaction takes place to form a film, which will significantly improve this problem. Air/solid interface synthesis is also a strategy to prepare large-scale single crystal films. This strategy can prepare ultrathin COF films with a thickness of several nanometers. Generally, surfactants are used to induce the regular arrangement of monomers on the surface of solvents. It usually takes a week to produce a better crystalline film.

Up to now, there are still few reports of the synthesis of wafer-scale films, and the methods are not necessarily universal. To obtain large-scale crystalline films, the reaction rate of monomers needs to be very slow. It usually takes a week. For imine COF films, it is very important to find a suitable reagent to inhibit the diffusion of monomers because of their high reactivity. In addition, surfactant is a good choice to induce crystallization of thin films. In future research, the efficiency of interfacial film formation and the crystallinity of the film are still important factors to be considered. Improving the existing methods, judging the catalyst activity in the reaction, and adjusting the diffusion rate are all promising research contents.

The currently reported COF linkage bonds mainly include boron-oxygen bonds, imide bonds, and carbon-carbon double bonds. COF films with higher reversibility of boron-oxygen bonds and imine bonds are reported the most. However, sp^2 carbon-carbon COF films have not been reported. This type of COF powder with a high degree of conjugate has been well applied in the field of photocatalysis. In future research, it is a good research direction to prepare this type of COF thin films by advanced interfacial methodologies. With its excellent light absorption ability and photoelectron migration ability, sp^2 carbon-carbon COF films can show excellent performance in seawater evaporation and photocatalysis.

Acknowledgements

T.Z. acknowledges the Excellent Youth Foundation of Zhejiang Province of China (Grant No. LR21E030001) and the National Natural Science Foundation of China (Grant No. 52005491 and No. 52003279).

Conflict of interest

The authors declare no conflict of interest.

Author details

Tao Zhang* and Yuxiang Zhao
Key Laboratory of Marine Materials and Related Technologies, Zhejiang Key Laboratory of Marine Materials and Protective Technologies, Ningbo Institute of Materials Technology and Engineering, Chinese Academy of Sciences, Ningbo, China

*Address all correspondence to: tzhang@nimte.ac.cn

References

[1] Côté AP, Benin AI, Ockwig NW, O'Keeffe M, Matzger AJ, Yaghi OM. Porous, crystalline, covalent organic frameworks. Science. 2005;**310**:1166-1170. DOI: 10.1126/science.1120411

[2] Evans AM, Parent LR, Flanders NC, Bisbey RP, Vitaku E, Kirschner MS, et al. Seeded growth of single-crystal two-dimensional covalent organic frameworks. Science. 2018;**361**:52-57. DOI: 10.1126/science.aar7883

[3] Liu Y, Wu H, Li R, Wang J, Kong Y, Guo Z, et al. MOF-COF "alloy" membranes for efficient propylene/propane separation. Advanced Materials. 2022;**2022**:e2201423. DOI: 10.1002/adma.202201423

[4] Fan H, Peng M, Strauss I, Mundstock A, Meng H, Caro J. MOF-in-COF molecular sieving membrane for selective hydrogen separation. Nature Communications. 2021;**12**:38. DOI: 10.1038/s41467-020-20298-7

[5] Gao H, Neale AR, Zhu Q, Bahri M, Wang X, Yang H, et al. A Pyrene-4,5,9,10-Tetraone-based covalent organic framework delivers high specific capacity as a Li-ion positive electrode. Journal of the American Chemical Society. 2022;**144**:9434-9442. DOI: 10.1021/jacs.2c02196

[6] Xu J, An S, Song X, Cao Y, Wang N, Qiu X, et al. Towards high performance Li-S batteries via sulfonate-rich COF-modified separator. Advanced Materials. 2021;**33**:e2105178. DOI: 10.1002/adma.202105178

[7] Zhao J, Ying Y, Wang G, Hu K, Yuan YD, Ye H, et al. Covalent organic framework film protected zinc anode for highly stable rechargeable aqueous zinc-ion batteries. Energy Stores. 2022;**48**:82-89. DOI: 10.1016/j.ensm.2022.02.054

[8] Hota MK, Chandra S, Lei Y, Xu X, Hedhili MN, Emwas AH, et al. Electrochemical thin-film transistors using covalent organic framework channel. Advanced Functional Materials. 2022;**32**:2201120. DOI: 10.1002/adfm.202201120

[9] Xu J, He Y, Bi S, Wang M, Yang P, Wu D, et al. An olefin-linked covalent organic framework as a flexible thin-film electrode for a high-performance Micro-supercapacitor. Angewandte Chemie, International Edition. 2019;**58**:12065-12069. DOI: 10.1002/anie.201905713

[10] Yu G, Zou X, Sun L, Liu B, Wang Z, Zhang P, et al. Constructing connected paths between UiO-66 and PIM-1 to improve membrane CO2 separation with crystal-like gas selectivity. Advanced Materials. 2019;**31**:e1806853. DOI: 10.1002/adma.201806853

[11] Rotter JM, Weinberger S, Kampmann J, Sick T, Shalom M, Bein T, et al. Covalent organic framework films through electrophoretic deposition—Creating efficient morphologies for catalysis. Chemistry of Materials. 2019;**31**:10008-10016. DOI: 10.1021/acs.chemmater.9b02286

[12] Wang L, Xu C, Zhang W, Zhang Q, Zhao M, Zeng C, et al. Electrocleavage synthesis of solution-processed, imine-linked, and crystalline covalent organic framework thin films. Journal of the American Chemical Society. 2022;**144**:8961-8968. DOI: 10.1021/jacs.1c13072

[13] Geng K, He T, Liu R, Dalapati S, Tan KT, Li Z, et al. Covalent organic

frameworks: Design, synthesis, and functions. Chemical Reviews. 2020;**120**:8814-8933. DOI: 10.1021/acs.chemrev.9b00550

[14] Wang H, Yang Y, Yuan X, Liang Teo W, Wu Y, Tang L, et al. Structure–performance correlation guided applications of covalent organic frameworks. Materials Today. 2022;**53**:106-133. DOI: 10.1016/j.mattod.2022.02.001

[15] Xu S, Richter M, Feng X. Vinylene-linked two-dimensional covalent organic frameworks: Synthesis and functions. Accounts of Chemical Research. 2021;**2**:252-265. DOI: 10.1021/accountsmr.1c00017

[16] Colson JW, Woll AR, Mukherjee A, Levendorf MP, Spitler EL, Shields VB, et al. Oriented 2D covalent organic framework thin films on single-layer graphene. Science. 2011;**332**:228-231. DOI: 10.1126/science.1202747

[17] Di M, Sun X, Hu L, Gao L, Liu J, Yan X, et al. Hollow COF selective layer based flexible composite membranes constructed by an integrated "casting-precipitation-evaporation" strategy. Advanced Functional Materials. 2022;**32**:2111594. DOI: 10.1002/adfm.202111594

[18] Chen L, Zhou C, Tan L, Zhou W, Shen H, Lu C, et al. Enhancement of compatibility between covalent organic framework and polyamide membrane via an interfacial bridging method: Toward highly efficient water purification. Journal of Materials Science. 2022;**656**:120590. DOI: 10.1016/j.memsci.2022.120590

[19] Cao L, Liu X, Shinde DB, Chen C, Chen IC, Li Z, et al. Oriented two-dimensional covalent organic framework membranes with high ion flux and smart gating nanofluidic transport. Angewandte Chemie, International Edition. 2022;**61**:e202113141. DOI: 10.1002/anie.202113141

[20] Sick T, Hufnagel AG, Kampmann J, Kondofersky I, Calik M, Rotter JM, et al. Oriented films of conjugated 2D covalent organic frameworks as photocathodes for water splitting. Journal of the American Chemical Society. 2018;**140**:2085-2092. DOI: 10.1021/jacs.7b06081

[21] Valentino L, Matsumoto M, Dichtel WR, Marinas BJ. Development and performance characterization of a Polyimine covalent organic framework thin-film composite nanofiltration membrane. Environmental Science & Technology. 2017;**51**:14352-14359. DOI: 10.1021/acs.est.7b04056

[22] Zhao Z, Chen W, Impeng S, Li M, Wang R, Liu Y, et al. Covalent organic framework-based ultrathin crystalline porous film: Manipulating uniformity of fluoride distribution for stabilizing lithium metal anode. Journal of Materials Chemistry A. 2020;**8**:3459-3467. DOI: 10.1039/c9ta13384d

[23] Dong R, Zhang T, Feng X. Interface-assisted synthesis of 2D materials: Trend and challenges. Chemical Reviews. 2018;**118**:6189-6235. DOI: 10.1021/acs.chemrev.8b00056

[24] Yuan S, Li X, Zhu J, Zhang G, Van Puyvelde P, Van der Bruggen B. Covalent organic frameworks for membrane separation. Chemical Society Reviews. 2019;**48**:2665-2681. DOI: 10.1039/c8cs00919h

[25] Sahabudeen H, Qi H, Glatz BA, Tranca D, Dong R, Hou Y, et al. Wafer-sized multifunctional polyimine-based two-dimensional conjugated polymers with high mechanical stiffness. Nature

Communications. 2016;7:13461. DOI: 10.1038/ncomms13461

[26] Matsumoto M, Dasari RR, Ji W, Feriante CH, Parker TC, Marder SR, et al. Rapid, low temperature formation of imine-linked covalent organic frameworks catalyzed by metal triflates. Journal of the American Chemical Society. 2017;**139**:4999-5002. DOI: 10.1021/jacs.7b01240

[27] Matsumoto M, Valentino L, Stiehl GM, Balch HB, Corcos AR, Wang F, et al. Lewis-acid-catalyzed interfacial polymerization of covalent organic framework films. Chem. 2018;**4**:308-317. DOI: 10.1016/j.chempr.2017.12.011

[28] Zhao S, Jiang C, Fan J, Hong S, Mei P, Yao R, et al. Hydrophilicity gradient in covalent organic frameworks for membrane distillation. Nature Materials. 2021;**20**:1551-1558. DOI: 10.1038/s41563-021-01052-w

[29] Zhong Y, Cheng B, Park C, Ray A, Brown S, Mujid F, et al. Wafer-scale synthesis of monolayer two-dimensional porphyrin polymers for hybrid superlattices. Science. 2019;**366**:1379-1384. DOI: 10.1126/science.aax9385

[30] Dey K, Pal M, Rout KC, Kunjattu HS, Das A, Mukherjee R, et al. Selective molecular separation by Interfacially crystallized covalent organic framework thin films. Journal of the American Chemical Society. 2017;**139**:13083-13091. DOI: 10.1021/jacs.7b06640

[31] Wang X, Shi B, Yang H, Guan J, Liang X, Fan C, et al. Assembling covalent organic framework membranes with superior ion exchange capacity. Nature Communications. 2022;**13**:1020. DOI: 10.1038/s41467-022-28643-8

[32] Feldblyum JI, McCreery CH, Andrews SC, Kurosawa T, Santos EJ, Duong V, et al. Few-layer, large-area, 2D covalent organic framework semiconductor thin films. Chemical Communications. 2015;**51**:13894-13897. DOI: 10.1039/c5cc04679c

[33] Zhang T, Qi H, Liao Z, Horev YD, Panes-Ruiz LA, Petkov PS, et al. Engineering crystalline quasi-two-dimensional polyaniline thin film with enhanced electrical and chemiresistive sensing performances. Nature Communications. 2019;**10**:4225. DOI: 10.1038/s41467-019-11921-3

[34] Liu K, Qi H, Dong R, Shivhare R, Addicoat M, Zhang T, et al. On-water surface synthesis of crystalline, few-layer two-dimensional polymers assisted by surfactant monolayers. Nature Chemistry. 2019;**11**:994-1000. DOI: 10.1038/s41557-019-0327-5

[35] Sahabudeen H, Qi H, Ballabio M, Položij M, Olthof S, Shivhare R, et al. Highly crystalline and semiconducting imine-based two-dimensional polymers enabled by interfacial synthesis. Angewandte Chemie, International Edition. 2020;**59**:6028-6036. DOI: 10.1002/anie.201915217

[36] Park S, Liao Z, Ibarlucea B, Qi H, Lin H-H, Becker D, et al. Two-dimensional Boronate Ester covalent organic framework thin films with large single crystalline domains for a neuromorphic memory device. Angewandte Chemie, International Edition. 2020;**59**:8218-8224. DOI: 10.1002/anie.201916595

[37] Hu X, Zhan Z, Zhang J, Hussain I, Tan B. Immobilized covalent triazine frameworks films as effective photocatalysts for hydrogen evolution reaction. Nature Communications. 2021;**12**:6596. DOI: 10.1038/s41467-021-26817-4

[38] Ou Z, Liang B, Liang Z, Tan F, Dong X, Gong L, et al. Oriented growth

of thin films of covalent organic frameworks with large single-crystalline domains on the water surface. Journal of the American Chemical Society. 2022;**144**:3233-3241. DOI: 10.1021/jacs.1c13195

[39] Zwaneveld NAA, Pawlak R, Abel M, Catalin D, Gigmes D, Bertin D, et al. Organized formation of 2D extended covalent organic frameworks at surfaces. Journal of the American Chemical Society. 2008;**130**:6678-6679. DOI: 10.1021/ja800906f

[40] Guan CZ, Wang D, Wan LJ. Construction and repair of highly ordered 2D covalent networks by chemical equilibrium regulation. Chemical Communications. 2012;**48**:2943-2945. DOI: 10.1039/c2cc16892h

[41] Liu XH, Guan CZ, Ding SY, Wang W, Yan HJ, Wang D, et al. On-surface synthesis of single-layered two-dimensional covalent organic frameworks via solid-vapor interface reactions. Journal of the American Chemical Society. 2013;**135**:10470-10474. DOI: 10.1021/ja403464h

[42] Liu M, Liu Y, Dong J, Bai Y, Gao W, Shang S, et al. Two-dimensional covalent organic framework films prepared on various substrates through vapor induced conversion. Nature Communications. 2022;**13**:1411. DOI: 10.1038/s41467-022-29050-9

[43] Mo YP, Liu XH, Wang D. Concentration-directed polymorphic surface covalent organic frameworks: Rhombus, parallelogram, and Kagome. ACS Nano. 2017;**11**:11694-11700. DOI: 10.1021/acsnano.7b06871

[44] Yue JY, Mo YP, Li SY, Dong WL, Chen T, Wang D. Simultaneous construction of two linkages for the on-surface synthesis of imine-boroxine hybrid covalent organic frameworks. Chemical Science. 2017;**8**:2169-2174. DOI: 10.1039/c6sc03590f

[45] Hao S, Zhang T, Fan S, Jia Z, Yang Y. Preparation of COF-TpPa1 membranes by chemical vapor deposition method for separation of dyes. Chemical Engineering Journal. 2021;**421**:129750. DOI: 10.1016/j.cej.2021.129750

[46] Medina DD, Rotter JM, Hu Y, Dogru M, Werner V, Auras F, et al. Room temperature synthesis of covalent-organic framework films through vapor-assisted conversion. Journal of the American Chemical Society. 2015;**137**:1016-1019. DOI: 10.1021/ja510895m

[47] Bisbey RP, DeBlase CR, Smith BJ, Dichtel WR. Two-dimensional covalent organic framework thin films grown in flow. Journal of the American Chemical Society. 2016;**138**:11433-11436. DOI: 10.1021/jacs.6b04669

[48] Wang F, Zhang Z, Shakir I, Yu C, Xu Y. 2D polymer nanosheets for membrane separation. Advancement of Science. 2022;**9**:e2103814. DOI: 10.1002/advs.202103814

[49] Li J, Jing X, Li Q, Li S, Gao X, Feng X, et al. Bulk COFs and COF nanosheets for electrochemical energy storage and conversion. Chemical Society Reviews. 2020;**49**:3565-3604. DOI: 10.1039/d0cs00017e

[50] Rodriguez-San-Miguel D, Montoro C, Zamora F. Covalent organic framework nanosheets: Preparation, properties and applications. Chemical Society Reviews. 2020;**49**:2291-2302. DOI: 10.1039/c9cs00890j

[51] Li G, Zhang K, Tsuru T. Two-dimensional covalent organic framework

(COF) membranes fabricated via the assembly of exfoliated COF nanosheets. ACS Applied Materials & Interfaces. 2017;**9**:8433-8436. DOI: 10.1021/acsami.6b15752

[52] Burke DW, Sun C, Castano I, Flanders NC, Evans AM, Vitaku E, et al. Acid exfoliation of imine-linked covalent organic frameworks enables solution processing into crystalline thin films. Angewandte Chemie, International Edition. 2020;**59**:5165-5171. DOI: 10.1002/anie.201913975

Chapter 2

Covalent Organic Frameworks for Ion Conduction

Fei Lu and Yanan Gao

Abstract

Covalent organic frameworks (COFs) are an emerging class of crystalline porous materials constructed by the precise reticulation of organic building blocks through dynamic covalent bonds. Due to their facile preparation, easy modulation and functionalization, COFs have been considered as a powerful platform for engineering molecular devices in various fields, such as catalysis, energy storage and conversion, sensing, and bioengineering. Particularly, the highly ordered pores in the backbones with controlled pore size, topology, and interface property provide ideal pathways for the long-term ion conduction. Herein, we summarized the latest progress of COFs as solid ion conductors in energy devices, especially lithium-based batteries and fuel cells. The design strategies and performance in terms of transporting lithium ions, protons, and hydroxide anions are systematically illustrated. Finally, the current challenges and future research directions on COFs in energy devices are proposed, laying the groundwork for greater achievements for this emerging material.

Keywords: COFs, ion conduction, lithium ion, proton, hydroxide

1. Introduction

The development of society depends on the effective use of new energy, which relies on the innovation of novel energy storage technology. Environmentally friendly energy storage devices such as lithium (Li)-ion batteries have achieved great success in the fields of consumer electronics and electric vehicles due to their excellent energy density. However, the large application of Li-ion batteries is limited by the use of liquid electrolytes, which suffer from the potential risk of leakage, flammability, and narrow voltage windows [1]. In contrast, solid polymer electrolytes with greater thermal and chemical stability have been considered as the promising candidates for applications in commercial energy devices including rechargeable batteries and fuel cells [2–4]. Recent representative solid-polymer electrolytes mostly involve with high-molecular-weight fluoro-containing polymers such as Nafion and polyolefin-type membranes [5]. However, the sever capacity fade is inevitable for these membranes due to the low ionic conductivity, especially under some extreme conditions including but not limited to high temperature and low relative humidity [6, 7]. Thus, solid-state electrolytes with outstanding ionic conductivity and superior stability are in high requirement.

The porous materials with high porosity and facial functionality, such as metal-organic frameworks (MOFs), polymers of intrinsic microporosity (PIMs), porous aromatic frameworks (PAFs), and covalent organic frameworks (COFs), offer potential high performance as solid electrolytes for energy devices. However, MOFs tend to decompose during battery cycling due to the low thermal and electrochemical stabilities. However, PIMs would reshaped their ultra-micropores to meso- and macropores under alkaline conditions, which reduces their electrochemical performance. PAFs with strong carbon-carbon bonds have a stable framework in harsh acidic and alkaline environments. But the synthesis methodologies of PAFs are very limited. Compared with all these porous materials, COFs can overcome the abovementioned disadvantages and therefore act as suitable candidates for energy device applications. COFs are synthesized by the polycondensation reaction to form covalent bonds between lightweight-atom-containing monomers, such as carbon (C), oxygen (O), nitrogen (N), hydrogen (H), boron (B), etc. [8, 9]. Most COFs are synthesized under thermodynamic control to modulate the reversibility of bond formation and breakage. As a result, highly periodic networks with defined pore size, shape, topology, and crystalline lattice will be formed according to the building blocks. The porosity generated by the geometries of monomers as well as the stable covalent bonds makes COFs preferable platforms for solid-state ion conduction. To be specific, 2D or 3D nanochannels can be constructed by the pores of COFs, through which the ions can transport. For the ions such as lithium ions (Li^+), protons (H^+), and hydroxides (OH^-), the conduction behavior mainly dominated by two mechanisms, which are hopping mechanism and vehicular mechanism [10–12]. For hopping mechanism, ions are preferable to hop between counter-charged adjacent sites with lower energy requirement to transport. While for vehicular mechanism, ions prefer bonding to some vehicular carriers such as H_2O (for H^+) or anions (for Li^+) to form large clusters, which needs higher energy to move. From this sense, the well-defined nanochannels of COFs, which could be further precisely installed ionic groups on the pore walls, provide an ideal pattern for ion conduction [13, 14]. In addition, the possibility to introduce extra charged species into the pores of COFs offers another opportunity for the high performance as ion conductors [15].

The superiority of COFs as high-performance ion conductors can also be described by the point of diffusion energy barrier of ionic migration [16]. For the liquid electrolytes, the charge carriers are surrounded by uniform and homogeneous solvents and thus can be quickly conducted by the exchange with solvating molecules, which

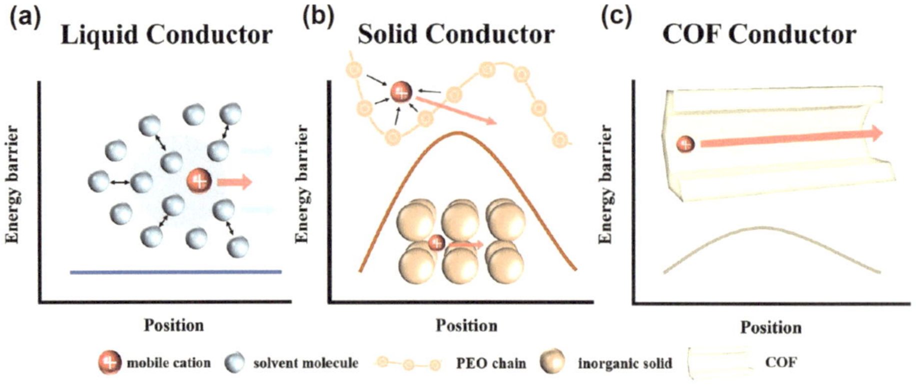

Figure 1.
Schematics of the ion-conducting behavior and diffusion energy barrier in (a) liquid conductor, (b) typical solid conductor, and (c) COF conductor, respectively. Reproduced with permission from Ref. [16]. Copyright 2022 Wiley-VCH GmbH.

generally produce high ionic conductivity. Thus, the diffusion energy barrier for ion conducting in liquid electrolytes can be considered flat (**Figure 1a**). However, the ion transporting in the solid-state conductors needs to overcome high energy barrier, which is related to the migration of charge carriers through the segmental motion of polymers or periodic crystalline space of inorganic solid (**Figure 1b**) [17, 18]. While for COFs, the charge species tend to migrate along the nanochannels due to the large free volume combined with the inside ionic sites, thus resulting in a lower diffusion energy barrier than that in typical solid-state conductor (**Figure 1c**). Thus, COFs are considered to be an excellent solid ion conductor.

In this context, we focus on the application of COFs as solid-state polyelectrolytes in energy devices, especially lithium-based batteries and fuel cells. The design strategies, nanostructures, and performance in terms of transporting Li^+, H^+, and OH^- are systematically illustrated. Finally, the current challenges and future research directions for the utilization of COFs in energy devices are proposed.

2. COFs for lithium-ion conduction

Li-ion batteries have been considered as the mainstream devices in commercial portable electronics. However, traditional Li-ion batteries suffer from serious safety risk due to the utilization of liquid electrolytes consisting of Li salts and flammable organic solvents. The long-term performance of Li-ion batteries is also limited due to the narrow voltage windows of liquid electrolytes. It means that the charging process would induce the decomposition of organic solvents and result in obvious capacity fade [19]. Additionally, the separators that are essential in liquid-electrolyte batteries commonly exhibit low conductivity, also inducing the decrease in the performance of Li-ion batteries [20]. Thus, the development of polymer electrolytes for all-solid-state batteries can address the above safety problems. The third key parameter is the Li^+ transference number (t_{Li+}), which represents the number of Li^+ transferred per Faraday of charge during charge and discharge process. The values of t_{Li+} for most polymer electrolytes are always lower than 0.5. These low t_{Li+} values would generate a ion concentration gradient between cathode and anode, leading to the polarization and limiting the charging/discharging rate and lifetime of batteries [3, 21]. Based on this background, the state-of-the-art polymer electrolytes should possess high ionic conductivity, high t_{Li+} values, and better electrochemical stability.

Due to the particular nanostructures and properties, COFs can act as an excellent Li^+ conductor due to some intrinsic merits. Firstly, the well-defined open channels of COFs provide fast pathways for the conduction of ions by reduced diffusion energy barrier. Second, the organic nature endows COFs' flexibility to introduce functional group to facilitate the dissociation of lithium salt and enhance the trapping of anions, which is beneficial to improve t_{Li+} value. Thirdly, the superior electrochemical and thermal stabilities make COFs available in the practical application.

In 2016, Zhang and coworkers synthesized a novel type of ionic COFs (ICOFs) containing sp^3 hybridized boron anions by the formation of spiroborate linkages [22]. After immobilization of Li^+ into the nanochannels, the obtained ICOF-2 exhibited an ionic conductivity of 3.05×10^{-5} S/cm and an average t_{Li+} value of 0.80 at room temperature, which was ascribed to the predesigned pathway for ion conduction. However, insertion of Li^+ into the bare nanochannels of COF can only produce a limited ionic conductivity. Inspired by the poly(ethylene oxide) (PEO)-based electrolytes, which can solvate Li^+ for fast ion conduction by the segmental motion, Jiang and coworkers firstly integrated the flexible oligo(ethylene oxide) chains on the pore walls of COFs [23]. Compared with the bare COF (Li^+@TPB-DMTP-COF), the EO-modified COF (Li^+@TPB-BMTP-COF) can exhibit an enhanced ionic conductivity and a lower energy barrier (**Figure 2a**). Furthermore, Horike *et al.* developed a bottom-up method to accumulate different concentration of glassy PEO moieties into the nanochannels of COF (**Figure 2b**) [24]. As a result, the material containing highest density accumulation of PEO (COF-PEO-9-Li) exhibited the superior ionic conductivity over 10^{-3} S/cm at 200°C. The similar trend was also demonstrated by branched PEO-functionalized COFs as solid conductors [26]. Recently, Horike used the similar side-chain engineering strategy to fabricate a gel-state COF (COF-Gel) with facile processability by implanting soft branched alkyl chains as internal plasticizers into the nanopores of COFs (**Figure 2c**) [25]. Benefiting from the gel-state morphology, the COF-Gel can be easily manufactured into gel electrolyte with controlled shape and thickness with enhanced ion conducting kinetics and reduced interfacial resistance, which showed a better performance in $LiFePO_4$-Li cell than the typical liquid electrolyte (**Figure 2d**).

The simple permeation of lithium salts into the nanopores of COFs usually results in a relatively low ionic conductivity and poor ion diffusion kinetics due to the closely associated ion pairs. While enhancing the binding interaction between the anions of lithium salt and COF backbones, the transfer of Li^+ would be promoted. In 2018, Chen et al. synthesized a cationic COF incorporated with LiTFSI to improve the ionic conductivity [27]. Compared with the neutral framework, the cationic skeleton can generate stronger dielectric screening to split Li^+ and the related anions, increasing the amount of free Li^+ and resulting in an improved solid-state ionic conductivity up to 2.09×10^{-4} S/cm at 70°C (**Figure 3a**). Similarly, Feng and coworkers employed imidazolium monomer as the building block to construct cationic COF to enhance the trapping of counter ions of lithium salts and improve the lithium ionic conductivity (**Figure 3b**) [28]. The obtained Im-COF-TFSI possessed the ionic conductivity as

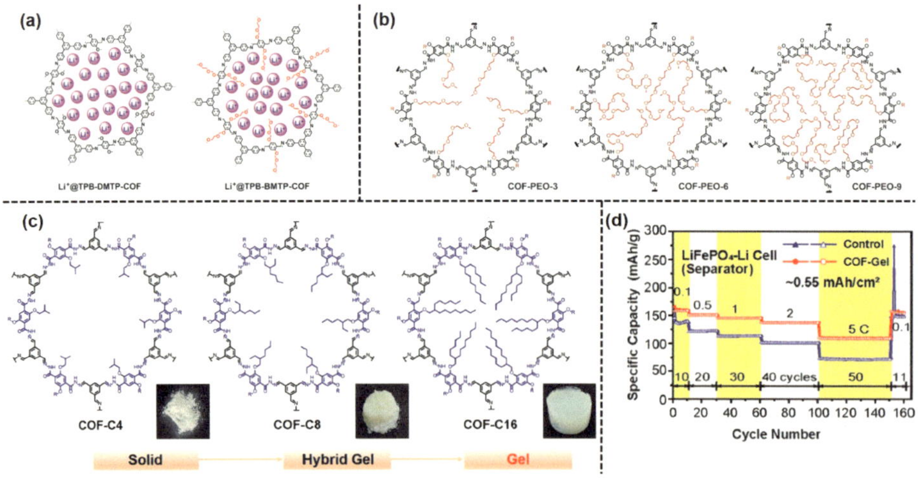

Figure 2.
(a) Structural representation of Li^+@TPB-DMTP-COF with bare pore walls and Li^+@TPB-DMTP-COF with EO-modified pore walls. Reproduced with permission from Ref. [23]. Copyright 2018 American Chemical Society. (b) Structural representation of COFs containing different concentration of PEO moieties. Reproduced with permission from Ref. [24]. Copyright 2018 American Chemical Society. (c) Structural representation and the corresponding state of COF with alkyl chains. (d) Rate performance of COF-Gel in a $LiFePO_4$-Li cell. Reproduced with permission from Ref. [25]. Copyright 2021 Wiley-VCH GmbH.

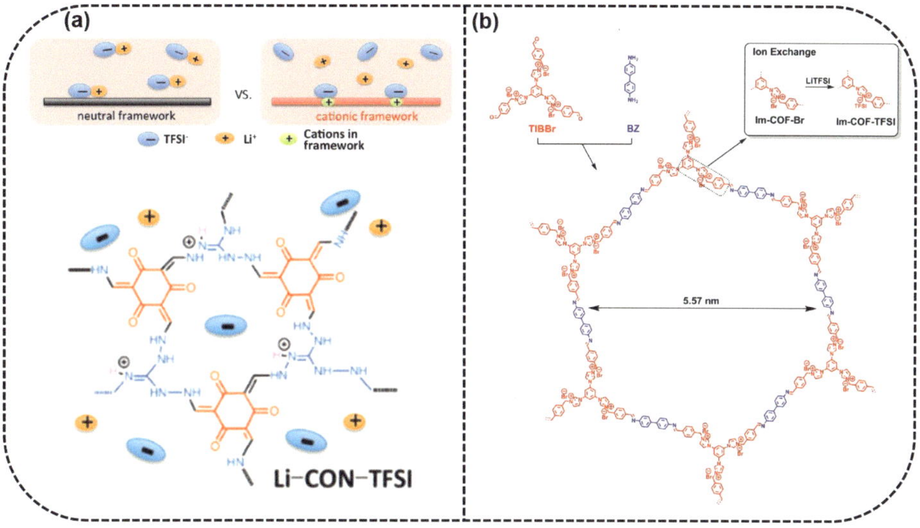

Figure 3.
(a) Structural representation of the cationic COF for binding with anions as Li^+ conductors. Reproduced with permission from Ref. [27]. Copyright 2018 American Chemical Society. (b) Synthesis of cationic Im-COF-TFSI. Reproduced with permission from Ref. [28]. Copyright 2020 Royal Society of Chemistry.

high as 4.64×10^{-4} S/cm at 80°C and 4.04×10^{-3} S/cm at 150°C, respectively. Recently, Han's group utilized the defects of COFs to introduce imidazolium groups onto the pore walls via the Schiff-base reaction [29]. After ion exchange with $TFSI^-$, the resultant dCOF-ImTFSI-Xs not only had the 2D pathways for Li^+ transport, but also contained the cationic moieties to promote the dissociation of lithium salts.

Although COFs have been demonstrated as a new concept of solid-state Li^+ conductors, most of the reported works are involved with the incorporation of lithium salts into the nanopores of COFs to facilitate ion transfer, thus failing to realize a single Li^+ conduction, which performs the t_{Li+} value close to unity (**Figure 4a**). To address this issue, Lee and coworkers presented a sulfonated COF (TpPa-SO_3Li) with anionic framework and well-defined directional channels (**Figure 4b**) [30]. As the sulfonates were covalently anchored to the pore walls, a real single Li^+ conductor can be achieved. The obtained TpPa-SO_3Li showed the t_{Li+} value as high as 0.9 at room temperature, thereby possessing a more stable Li plating/stripping on lithium anode than the previously reported COF-based ion conductors (**Figure 4c**). Recently, Loh et al. used a post-synthetic method to graft sulfonate groups on the pore wall of COFs and exfoliated COF powders to nanosheets to create better ion conduction channels because the repulsive forces between Li^+ were stronger than the van der Waals interactions between frameworks (**Figure 4d**) [31]. As a result, the sulfonated COF nanosheets achieved the ionic conductivity of 0.9×10^{-5} S/cm at −40°C and 1.17×10^{-4} S/cm at 100°C with t_{Li+} value of 0.92 when equipped in an all-solid-state battery, indicating the possibility to be practically applied under harsh conditions.

3. COFs for proton conduction

Due to the high energy conversion efficiency, low emission, fuel flexibility, and mild operation accessibility, fuel cells are considered as electrochemical power plants,

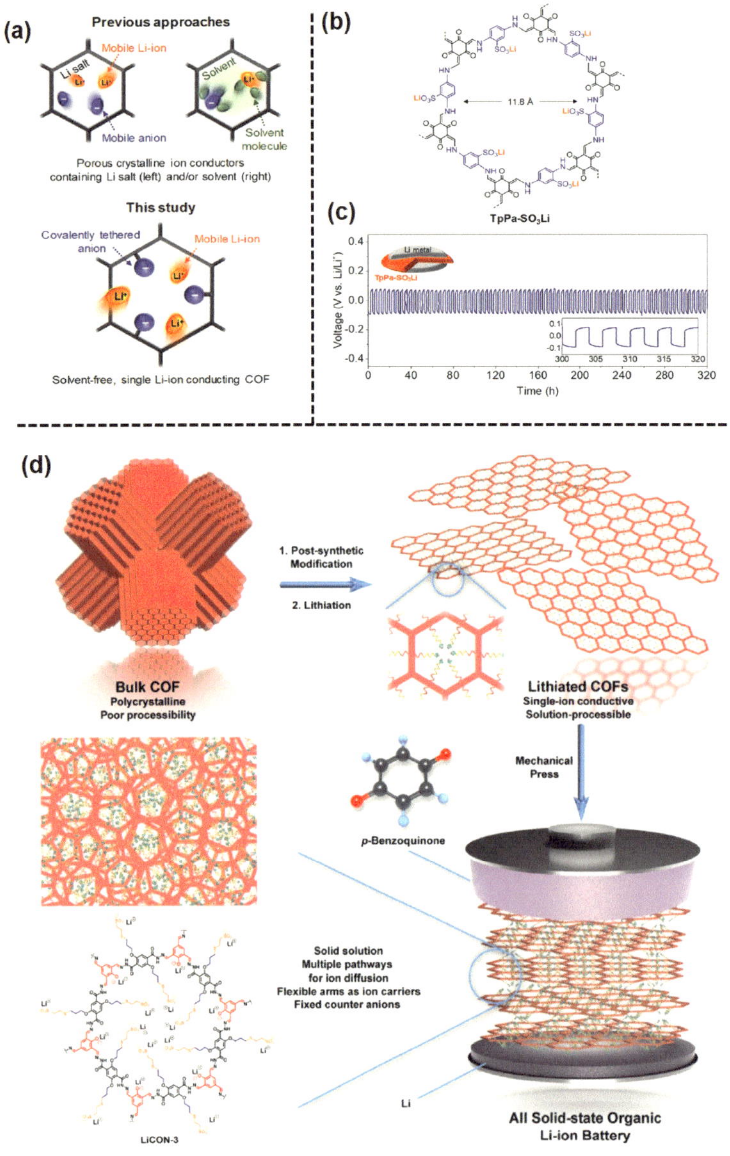

Figure 4.
(a) Conceptual illustrations of ion transport phenomena in the porous crystalline ion conductors: previous approaches (top) and this study (bottom). (b) Chemical structure of lithium sulfonated COF (TpPa-SO_3Li). (c) Galvanostatic Li plating/stripping profile of the Li/Li symmetric cell containing TpPa-SO_3Li. Reproduced with permission from Ref. [30]. Copyright 2019 American Chemical Society. (d) Schematic illustration describing the fabrication of all-solid-state organic Li-ion battery using lithiated COF nanosheets. Reproduced with permission from Ref. [31]. Copyright 2020 American Chemical Society.

which convert chemical energy to electrical energy by the cost of specific fuels. Among the typical fuel cells, which differentiated by the type of electrolytes (phosphoric acid, proton exchange membrane, oxide, and alkaline), the proton exchange membrane fuel cell (PEMFC) has attracted the most attentions due to the relatively lower operation temperature [32]. Except for PEMFC, proton-conducting materials are also the key component for other electrochemical devices such as supercapacitors, proton sieving, proton transistors and hydrogen sensors [33]. Up to now, the universally utilized proton-conducting membranes are sulfonated polymers such as Nafion. Nafion can display proton conductivity as high as 10^{-1} S/cm in fully hydrated state at a moderate temperature [5]. However, when the temperature exceeds 100°C, the proton conductivity would dramatically decrease due to the loss of water. Other polymer-based proton conducting materials such as polybenzimidazole (PBI)-based membranes can perform excellent properties in 150–200°C. While the heterogeneous random structures and the amorphous nature hinder the efficient analysis of conducting mechanism and determination of structure-property relationship at the molecular level [34]. In this regard, COFs with precise structural designability, synthetic controllability, and available functionality hold the promise to work as superior proton conductors.

Inspired by the structure of Nafion, sulfonated COFs would be the ideal candidates for proton conduction and fuel cell practical application. In 2016, Banerjee's group firstly synthesized sulfonic-acid-based COFs (TpPa-SO_3H) by a Schiff-base reaction of 1,3,5-triformylphloroglucinol and 2,5-diaminobenzenesulfonic acid to form periodic intervals with free -SO_3H groups within COF backbones to promote the proton hopping along the hexagonal 1D channels [35]. The intrinsic proton conductivity of TpPa-SO_3H can reach 1.7×10^{-5} S/cm at 120°C under anhydrous condition. After shortly, Zhao et al. reported two sulfonated COFs, NUS-9(G) and NUS-10 (G), by liquid-assisted grinding strategy at room temperature (**Figure 5a**) [36]. The obtained NUS-9(G) and NUS-10 (G) exhibited hexagonal architecture and displayed eclipsed AA stacking layers (**Figure 5b**). Due to the pre-implanted free -SO_3H groups, NUS-9(G) showed a proton conductivity of 1.5×10^{-4} S/cm at room temperature and 33% relative humidity (RH), which was increased to 3.96×10^{-2} S/cm at 97% RH. While for NUS-10 (G), which possessed twice as many free -SO_3H groups as NUS-9(G), the intrinsic proton conductivity increased to 2.8×10^{-4} S/cm at 33% RH and 3.96×10^{-2} S/cm at 97% RH with long-term stability. Very recently, Zhu's group used surface-initiated condensation polymerization to synthesize the same sulfonic COF TpPa-SO_3H (**Figure 5c**) [37]. Through the precise control of polymerization time, the thickness of SCOF layer can be tuned from 10 to 100 nm, overcoming the processable challenge of COFs. The obtained free-standing COF membrane exhibited a proton conductivity of 0.54 S/cm at 80°C under fully hydrated state.

For these sulfonated COFs, the high water-assisted proton conductivity could be attributed to the presence of aligned -SO_3H groups on the pore walls of COFs, which not only enhance the adsorption of water but also facilitate the formation of hydrophilic domains to generate proton conducting pathways. At a high RH percentage, the proton transportation is mainly dominated by the hopping mechanism benefiting from the continuous hydrogen bonds between H_2O and -SO_3H groups. However, under low humidity condition, the significant decrease of proton conductivity would be still inevitable, which is similar to Nafion. Otherwise, the preinstallation of proton conducting groups onto the pore wall of COFs is sometimes difficult. Thus, incorporation of guest protonic species into the nanochannels of COFs would provide a more effective approach to improve the proton conductivity.

In one aspect, some proton donors can incorporate with COFs to trigger proton conductivity by donating more protons or facilitating the formation of hydrogen bonds. In 2016, Jiang and coworkers developed a highly robust COF, TPB-DMTP-COF, with hexagonally aligned, dense, mesoporous channels (**Figure 6a**) [38]. By loading the N-heterocyclic proton carriers, 1,2,4-triazole (trz) and imidazole (im), the anhydrous proton conductivity of trz@TPB-DMTP-COF and im@TPB-DMTP-COF can reach the maximum of 1.1×10^{-3} S/cm and 4.37×10^{-3} S/cm at 130°C,

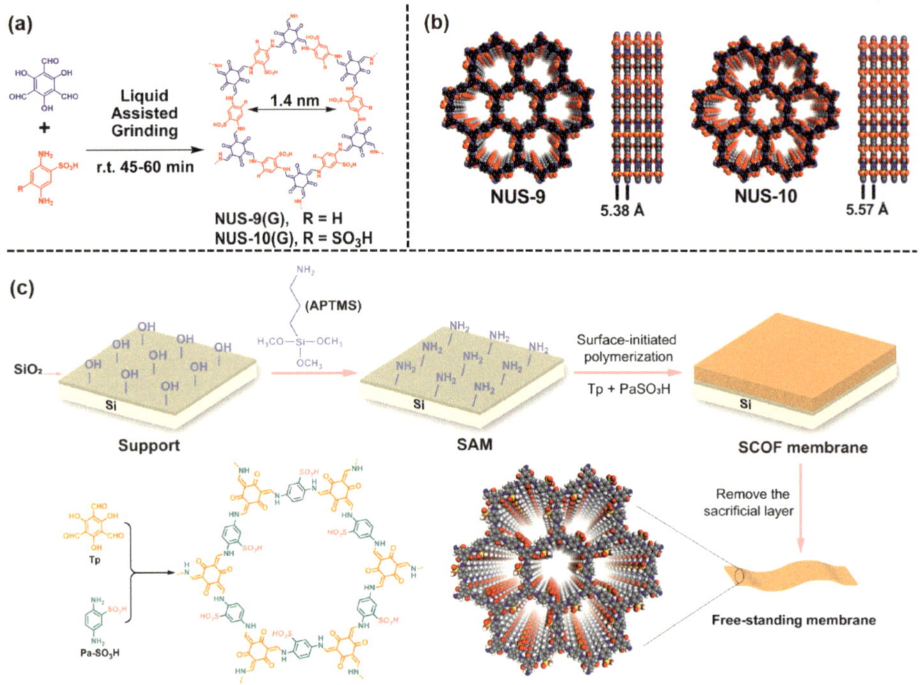

Figure 5.
(a) Synthetic scheme of NUS-9(G) and NUS-10(G) via liquid-assisted grinding at room temperature. (b) Crystal structure and representation of the eclipsed AA stacking for NUS-9(G) and NUS-10(G), respectively. Reproduced with permission from Ref. [36]. Copyright 2016 American Chemical Society. (c) Preparation of the SCOF membrane grafted on silicon wafers and the molecular structures TpPa-SO_3H. Reproduced with permission from Ref. [37]. Copyright 2021 Wiley-VCH GmbH.

respectively. The activation energy calculation also demonstrated that protons were transported by hopping along the interconnected hydrogen bonding networks of proton carriers.

Besides, phosphoric acid (H_3PO_4) is also a good proton donors for the extrinsic incorporation with COFs [40]. In 2020, Horike's group reported perfluoroalkyl-functionalized COFs (COF-Fx-H) with super hydrophobic well-defined 1D channels to accommodate a large amount of H_3PO_4 (**Figure 6b**) [39]. Due to the interactions between H_3PO_4 and the NH groups of framework as well as the fluorinated side chains, the guest H_3PO_4 could be anchored onto the pore walls through P=O...H−N, OH... N=C, and O−H...F−C hydrogen bonding networks, which further generated an efficient proton conducting pathway (**Figure 6c**). After 62 wt% loading of H_3PO_4, the maximum anhydrous proton conductivity reached 4.2×10^{-2} S/cm. Recently, Jiang's group also designed polybenzimidazole COFs in conjunction with H_3PO_4 to achieve stable and ultrafast proton conduction over a wide range of temperature [41]. Due to the presence of imine linkage and the benzimidazole chains, H_3PO_4 could be tightly locked by the electrostatic and hydrogen binding interactions. More importantly, the N atom of benzimidazole moieties could be protonated by H_3PO_4 and release open H_2PO4^- anion. Thus the proton conduction would be facilitated by the activated proton networks. As a result, the H_3PO_4@TPB-DABI-COF realized a hydrous proton conductivity of 8.35×10^{-3} S/cm at 160°C.

Although great progress has been achieved for improving the proton conductivity by extrinsic incorporation with proton carriers such as acids and N-heterocycles,

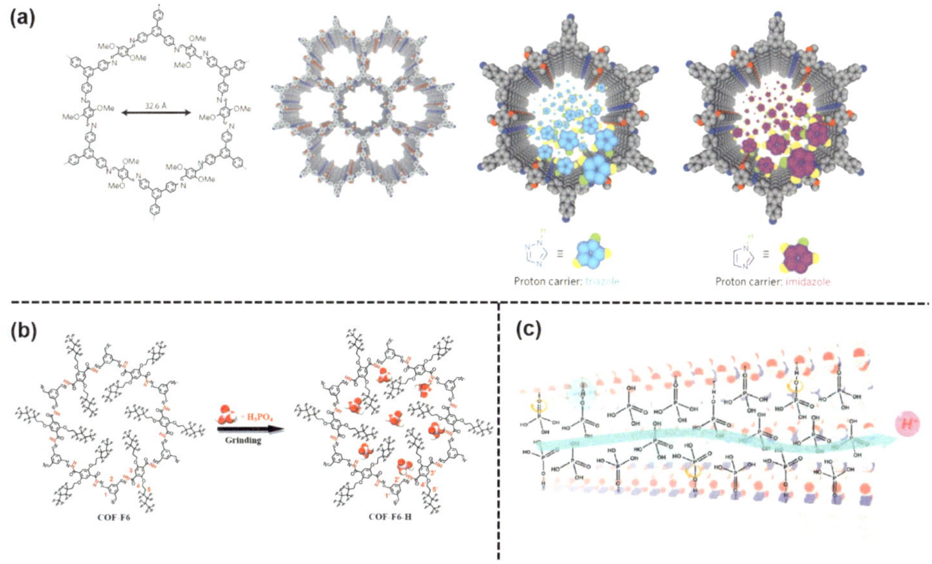

Figure 6.
(a) Chemical structure and hexagonal structure of TPB-DMTP-COF and graphic representation of 1,2,4-triazole and imidazole in the channels. Reproduced with permission from Ref. [38]. Copyright 2016 Nature Publishing Group. (b) Synthesis route of COF-Fx-H. (c) Illustration of proposed proton conducting mechanism. Reproduced with permission from Ref. [39]. Copyright 2020 American Chemical Society.

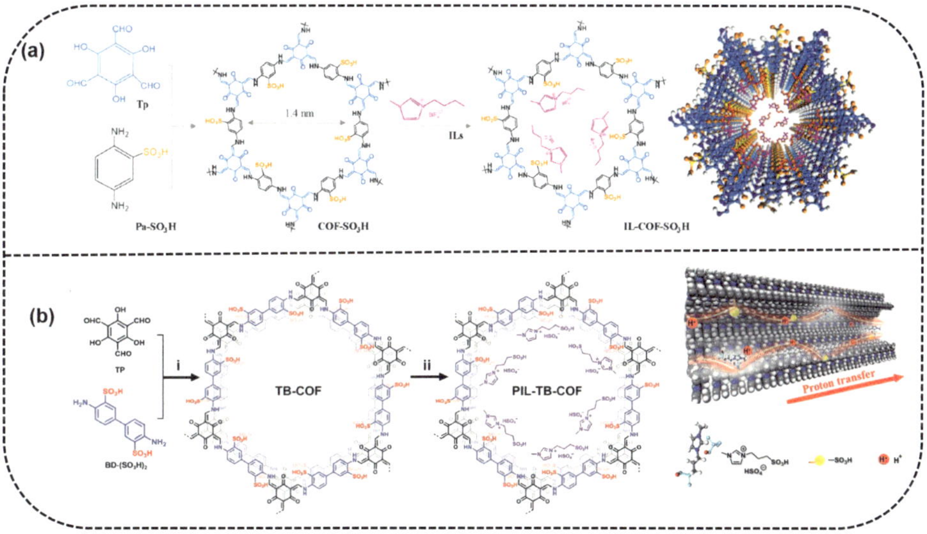

Figure 7.
(a) Schematic of the synthesis and structure of IL-COF-SO$_3$H. Reproduced with permission from Ref. [42]. Copyright 2021 Elsevier B. V. (b) Structure and illustration of the proton transfer of PIL-TB-COF. Reproduced with permission from Ref. [43]. Copyright 2022 The Royal Society of Chemistry.

relatively less attention has been paid on the proton conducting property of ionic liquids (ILs) impregnated COFs. In 2021, Tang and coworkers firstly reported an IL impregnated sulfonic-acid-based COF (IL-COF-SO$_3$H), which further combined with silk nanofibrils (SNFs) to fabricate a composite membrane (**Figure 7a**) [42]. The electrostatic interactions between imidazolium anions and sulfonic acids promoted the deprotonation to release more protons and immobilized ILs. The uniform distribution of ILs in the channels of COF-SO_3H could provide a large amount of hopping sites for protons. Moreover, the hydrogen bonding networks between SNFs and IL-COF-SO_3H could provide additional proton conduction pathways. Particularly, the IL-COF-SO_3H@SNF-35, which loaded 35 wt% SNFs, acquired an ionic conductivity of 224 mS/cm at 90°C and 100% RH. Very recently, Yan's group developed a protic ionic liquid (PIL), 1-methyl-3-(3-sulfopropyl) imidazolium hydrogensulphate ([PSMIm][HSO_4]) to incorporate with a high-density $-SO_3H$ functionalized COF (TB-COF) for efficient anhydrous proton conduction (**Figure 7b**) [43]. As expected, the addition of PIL into the nanochannels of COFs can significant increase the ionic conductivity from 1.52×10^{-4} S/cm to 2.21×10^{-3} S/cm at 120°C due to the increase of hopping sites for protons.

4. COFs for hydroxide anion conduction

Compared with PEMFC, alkaline fuel cells, which are operated on hydroxide anion transport, have attracted increasing attention due to the high energy density, rapid reaction kinetics, and low-cost catalyst [44]. As one of the critical components, the hydroxide conducting membrane affords the transfer of anions and determines the terminal electrochemical output. However, the high-performance hydroxide anion conduction is challenging because of its lower diffusion coefficient compared with protons [45]. Similar to proton conducting materials, the typical anion conducting membranes depend on polymer system, which can form percolated water channels via the microphase separation of hydrophobic/hydrophilic domains [46–48]. Multiple factors including the polarity of segments, distribution of charged moieties, and anion concentration have influence on the physical phase-separation process. Thus, precise control of the phase-segregated morphologies usually has of a large difficulty. Based on this background, COFs with structural tunability and functional pore surface build a powerful platform to achieve fast anion transport.

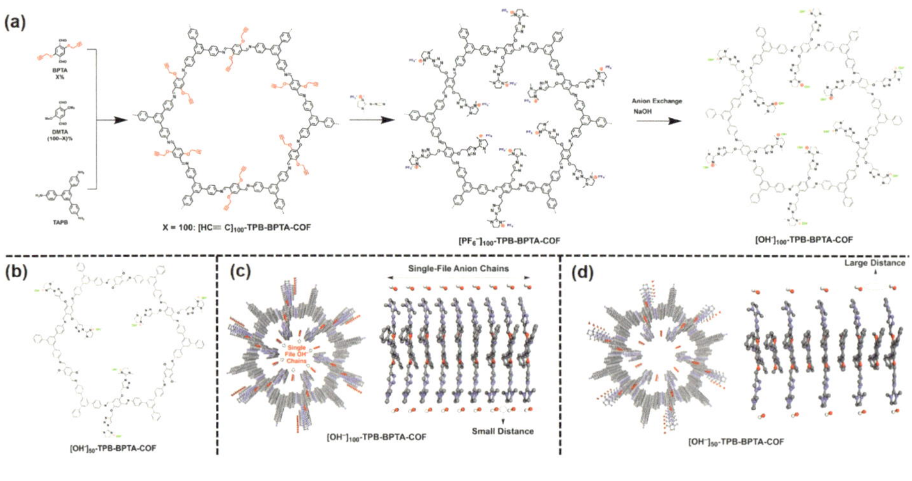

Figure 8.
(a) Synthesis route of [OH^-]$_{100}$-TPB-BPTA-COF. (b) Structure of [OH^-]$_{50}$-TPB-BPTA-COF. Reconstructed structures of (c) [OH^-]$_{100}$-TPB-BPTA-COF and (d) [OH^-]$_{50}$-TPB-BPTA-COF (red O, blue N, gray C, white H). Reproduced with permission from Ref. [49]. Copyright 2021 American Chemical Society.

For example, Jiang and coworkers constructed an anion-surfaced channels for hydroxide anion conduction via supramolecular self-assembly while maintaining both the ordering topology and skeleton stability of COFs [49]. The precursor COF with ethynyl side groups was synthesized by the condensation of C_3-symmetric 1,3,5-tris(4-aminophenyl)benzene (TAPB) as knot and 2,5-bis(2-propynyloxy)terephthalaldehyde (BPTA) and 2,5-dimethoxyterephthalaldehyde at different molar ratios (DMTA) as linker (**Figure 8a**). Then an azide-imidazolium salt was introduced into the pores by the click reaction with the ethynyl sides. After anion exchange, the hydroxide anion conducting $[OH^-]_{100}$-TPB-BPTA-COF was constructed (**Figure 8a**). The crystal structural analysis demonstrated that the imidazolium cations were extruded from pore walls and concentrated in the channel center aligning with OH^- at the end of cationic chains, thus creating a continuous anionic phase (**Figure 8c**). While for $[OH^-]_{50}$-TPB-BPTA-COF in which half of the edge units were appended with ethynyl groups (**Figure 8b**), the hydroxide anion interface was not continuous due to the reduced anion density (**Figure 8d**). Consequently, the conductivity of $[OH^-]_{100}$-TPB-BPTA-COF was 2–8 times higher than that of $[OH^-]_{50}$-TPB-BPTA-COF.

The poor processability of COFs generally produces insoluble powders, which dramatically limits their practical application as free-standing membranes. Thus, it is highly desirable to develop efficient methods to fabricate COF-based anion conducting membranes. Recently, Jiang's group has made remarkable achievements on engineering COF membranes via interfacial polymerization strategy [50–52].

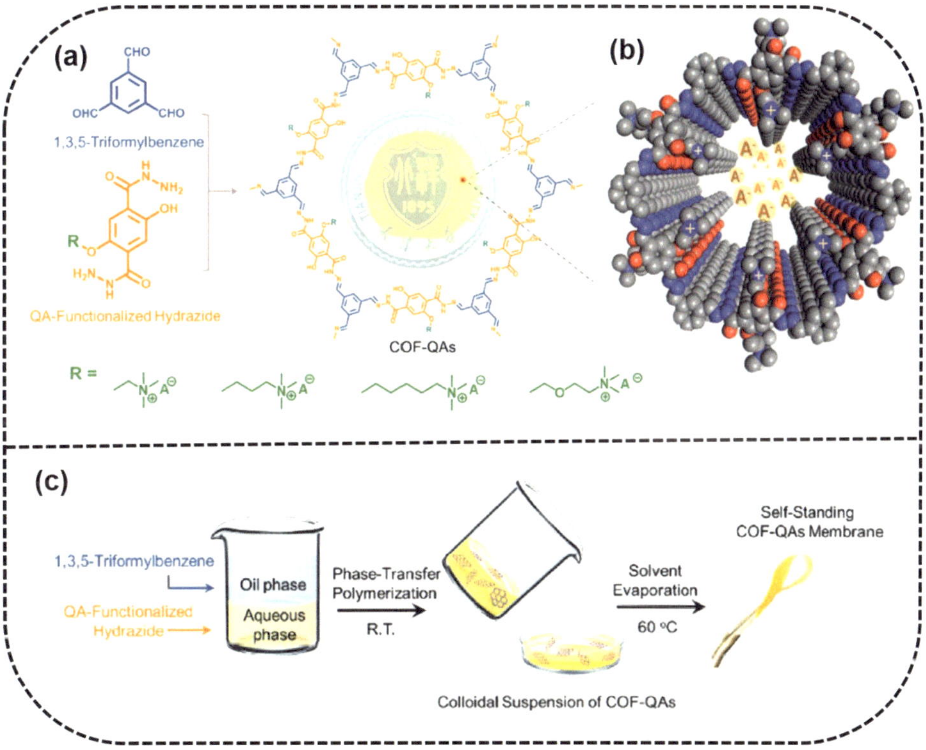

Figure 9.
(a) Scheme for the synthesis of COF-QAs. (b) Schematic of anion transport through the 1D channel of COF-QAs. (c) Schematic of COF-QAs membrane fabrication process. Reproduced with permission from Ref. [53]. Copyright 2020 WILEY-VCH Verlag GmbH & Co. KGaA, Weinheim.

For instance, they used hydrazide units functionalized with quaternary ammonium (QA) groups bearing different length of alkyl chains and aldehyde units to construct four quaternized COFs (COF-QAs) (**Figure 9a**) [53]. The well-defined ordered nanochannels with aligned QA cations provided an ultrafast pathway for anion transport (**Figure 9b**). In order to realize the self-standing membrane with robust crystalline framework, a phase-transfer polymerization process, which involved phase transfer of 1,3,5-triformylbenzene to polymerize with QA-functionalized hydrazide in a mesitylene-water system (**Figure 9c**). Due to the slight solubility in water, the aldehyde units would gradually diffuse from organic phase to aqueous solution to react with hydrazides, resulting in a stable colloidal suspension. Upon solvent removal, the COF nanoplates could further be assembled into free-standing membrane with identical crystalline structures, which exhibited the hydroxide conductivity as high as 212 mS/cm at 80°C. Very recently, they developed six QA-functionalized COFs via the assembly of hydrazides and aldehyde precursors by interfacial polymerization to systematically elucidate the impact of aldehyde size, electrophilicity, and hydrophilicity on the synthesis process as well as the anion conducting property of COFs [54]. Particularly, more hydrophilic aldehydes were preferable to react with hydrazides in the aqueous solution rather than the interface region, which led to the tight membrane. Compared with the loose membranes, the anion conductivity could improve around 4–8 times.

5. COFs for other ion conduction

Among various electrical devices to date, sodium-ion batteries have gained considerable attention due to its low cost and sustainability. Similar with lithium-ion batteries, sodium-ion batteries also suffer from the easy formation of dendrite in traditional liquid electrolyte and have high desire to develop solid ion conductors. To tackle these bottlenecks, Sun and coworkers studied the first example of carboxylic acid sodium functionalized COF (NaOOC-COF) as quasi-solid-state electrolyte to accelerate the transporting of Na^+ and simultaneously restrain the dendrite growth (**Figure 10a**). The covalently tethered carboxylic acid sodium groups in the pore wall of COFs provided sufficient content of Na^+ and favorable nanostructures for Na^+ migration. Benefiting from the well-defined ion channels, NaOOC-COF displayed an excellent conductivity of 2.68×10^{-4} S/cm at room temperature and high transference number of 0.9. Finally, NaOOC-COF devoted to durable cycling performance of Na plating/stripping and outstanding performance in solid-state battery [55].

Aqueous Zn-ion batteries are also a great promising energy storage system owing to the high energy density driven by multielectron redox ($Zn^{0/2+}$) and prominent safety supported by water-based electrolytes. However, the practical application has still been limited due to the lack of suitable electrolytes to ensure stable interface with electrodes. Recently, Lee and coworkers demonstrated for the first time to use COF-based single Zn^{2+} conductors, which can both secure interfacial stability with electrodes and exhibit competitive ionic conductivity [56]. A zinc sulfonated COF (TpPa-$SO_3Zn_{0.5}$, **Figure 10b**) with well-defined directional channels in which covalently anchored and delocalized sulfonates was designed to realize single Zn^{2+} conduction. From the molecular dynamics (MD) simulations, a significantly uniform Zn^{2+} flux was observed due to the anionic groups along the directional pores (**Figure 10c**). While in the control model of liquid electrolyte (LE), which is 2 M $ZnSO_4$ in H_2O, only randomly spread Zn^{2+} clusters can be observed due to the freely

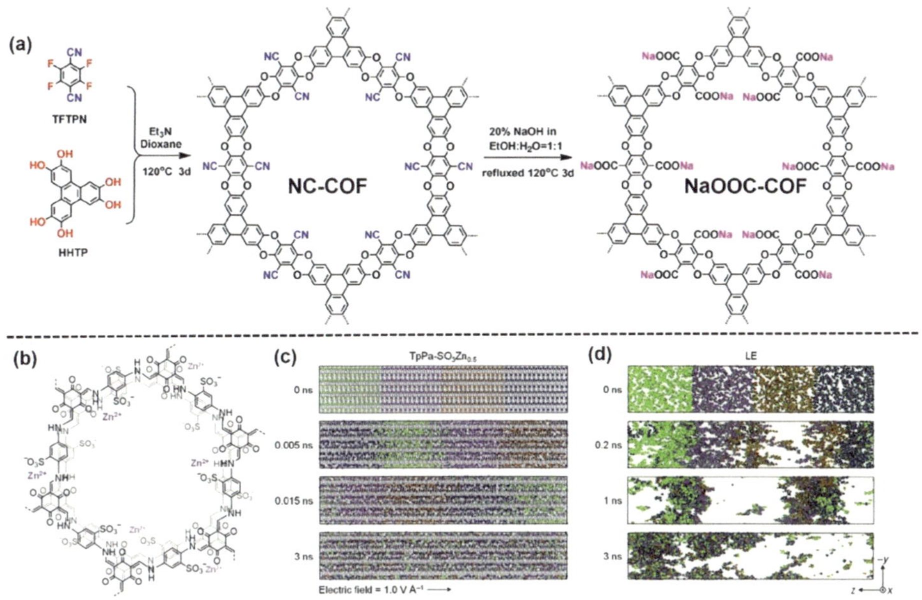

Figure 10.
(a) The synthetic route of NaOOC-COF. Reproduced with permission from ref. 55. Copyright 2021 Elsevier Ltd. (b) Chemical structure of TpPa-$SO_3Zn_{0.5}$. Representative snapshots obtained from the MD simulations showing time-dependent ion distributions in (c) TpPa-$SO_3Zn_{0.5}$ and (d) LE. Zn^{2+}: colored diversely for a clear representation of the movement, TpPa-SO_3^-: gray, SO_4^{2-}: green, H_2O: omitted for clarity. Reproduced with permission from Ref. [56]. Copyright 2020 The Royal Society of Chemistry.

mobile SO_4^{2-} (**Figure 10d**). As a result, TpPa-$SO_3Zn_{0.5}$ enabled the Zn-MnO_2 cells to exhibit a long-term cycling performance.

6. Conclusion and outlook

In this chapter, we summarized the recent progress of COFs as solid-state ion conductors in energy devices, especially lithium-based batteries and fuel cells. As the emerging crystalline porous materials with controllable chemistry, tunable topology, and well-defined order channels, COFs exhibit a promising performance to conduct lithium ion, proton, and hydroxide anion. However, the development of COF-based solid ion conductors is still in its infancy, and many challenges remain to be issued.

Firstly, most ion-conducting COFs relate to ionic frameworks. Compared with neutral COFs, the examples of ionic COFs are still limited due to the more restricted synthesis conditions for crystallization and ionization. Although a series of covalent bonds have been successfully applied to construct COFs, only a few of linkages afford the formation of ionic COFs. To date, most ionic COFs are formed by the imine bonds. Thus, deep chemistry insight and novel synthetic approach to ionize COFs are in high demand. In addition, universal strategies to construct 3D ionic COFs, which are scientifically intriguing with unique properties, are also requiring since most COFs have 2D frameworks.

Secondly, most COFs are synthesized via a solvothermal method under harsh reaction conditions with powder products. Thus, the large-scale synthesis of COFs at industrial level is still challenging. To achieve more practical application, the large-scale synthesis with retaining the crystallinity and porosity of COFs is of critical significance. Moreover, the powder nature also hinders their application in electronic devices. Although some strategies such as interfacial polymerization can develop free-standing COF membranes to some extent, the limited mechanical property is always hard to meet the practical requirement. Therefore, efficient approaches to prepare COF membrane with good mechanical stability should be explored to enhance their practicality, especially in flexible electronic devices.

Thirdly, the development of COFs as ion conductors is in the initial state with most research interests focusing on the improvement of apparent performance via experimental investigations. To better clarify the structure-property relationship and guide the structural design, the theoretical simulations, which can provide more thorough insight on the ion transport mechanism in COFs, should be probed.

To sum up, COFs offer new opportunities for the solid ion conductors and exhibit tremendous advantages over other materials such as highly ordered pores, tailorable pore surface, tunable chemical composition, etc. Benefiting from the rapid development of experimental and theoretical tools, the electrochemical performance of COFs is expected to gain greater achievements.

Acknowledgements

The authors acknowledge financial support from the National Natural Science Foundation of China (21965011 and 21902092) and the Major Science and Technology Plan of Hainan Province (ZDKJ202016).

Conflict of interest

The authors declare no conflict of interest.

Author details

Fei Lu and Yanan Gao*
Key Laboratory of Ministry of Education for Advanced Materials in Tropical Island Resources, Hainan University, Haikou, China

*Address all correspondence to: ygao@hainanu.edu.cn

References

[1] Yamada Y, Wang J, Ko S, Watanabe E, Yamada A. Advances and issues in developing salt-concentrated battery electrolytes. Nature Energy. 2019;**4**:269-280. DOI: 10.1038/s41560-019-0336-z

[2] Zhang H, Li C, Piszcz M, Coya E, Rojo T, Rodriguez-Martinez LM, et al. Single lithium-ion conducting solid polymer electrolytes: Advances and perspectives. Chemical Society Reviews. 2017;**46**:797-815. DOI: 10.1039/c6cs00491a

[3] Zhu J, Zhang Z, Zhao S, Westover AS, Belharouak I, Cao PF. Single-ion conducting polymer electrolytes for solid-state lithium–metal batteries: Design, performance, and challenges. Advanced Energy Materials. 2021;**11**:2003836. DOI: 10.1002/aenm.202003836

[4] Min J, Barpuzary D, Ham H, Kang GC, Park MJ. Charged block copolymers: From fundamentals to electromechanical applications. Accounts of Chemical Research. 2021;**54**:4024-4035. DOI: 10.1021/acs.accounts.1c00423

[5] Mauritz KA, Moore RB. State of understanding of Nafion. Chemical Reviews. 2004;**104**:4535-4585. DOI: 10.1021/cr0207123

[6] Zhang H, Shen PK. Recent development of polymer electrolyte membranes for fuel cells. Chemical Reviews. 2012;**112**:2780-2832. DOI: 10.1021/cr200035s

[7] Shin DW, Guiver MD, Lee YM. Hydrocarbon-based polymer electrolyte membranes: Importance of morphology on ion transport and membrane stability. Chemical Reviews. 2017;**117**:4759-4805. DOI: 10.1021/acs.chemrev.6b00586

[8] Li Z, He T, Gong Y, Jiang D. Covalent organic frameworks: Pore design and Interface engineering. Accounts of Chemical Research. 2020;**53**:1672-1685. DOI: 10.1021/acs.accounts.0c00386

[9] Li Y, Chen W, Xing G, Jiang D, Chen L. New synthetic strategies toward covalent organic frameworks. Chemical Society Reviews. 2020;**49**:2852-2868. DOI: 10.1039/d0cs00199f

[10] Zhu H, Rana U, Ranganathan V, Jin L, O'Dell LA, MacFarlane DR, et al. Proton transport behaviour and molecular dynamics in the guanidinium triflate solid and its mixtures with triflic acid. Journal of Materials Chemistry A. 2014;**2**:681-691. DOI: 10.1039/c3ta13344c

[11] Pan J, Chen C, Li Y, Wang L, Tan L, Li G, et al. Constructing ionic highway in alkaline polymer electrolytes. Energy & Environmental Science. 2014;**7**:354-360. DOI: 10.1039/c3ee43275k

[12] Castiglione F, Ragg E, Mele A, Appetecchi GB, Montanino M, Passerini S. Molecular environment and enhanced diffusivity of Li+ ions in Lithium-salt-doped ionic liquid electrolytes. The Journal of Physical Chemistry Letters. 2011;**2**:153-157. DOI: 10.1021/jz101516c

[13] Liang X, Tian Y, Yuan Y, Kim Y. Ionic covalent organic frameworks for energy devices. Advanced Materials. 2021;**33**:e2105647. DOI: 10.1002/adma.202105647

[14] Zhang P, Wang Z, Cheng P, Chen Y, Zhang Z. Design and application of ionic covalent organic frameworks. Coordination Chemistry Reviews. 2021;**438**:213873. DOI: 10.1016/j.ccr.2021.213873

[15] Zhang Y, Wu MX, Zhou G, Wang XH, Liu X. A rising star from two worlds: Collaboration of COFs and ILs. Advanced Functional Materials. 2021;**31**:2104996. DOI: 10.1002/adfm.202104996

[16] Cao Y, Wang M, Wang H, Han C, Pan F, Sun J. Covalent organic framework for rechargeable batteries: Mechanisms and properties of ionic conduction. Advanced Energy Materials. 2022;**12**:2200057. DOI: 10.1002/aenm.202200057

[17] Osada I, de Vries H, Scrosati B, Passerini S. Ionic-liquid-based polymer electrolytes for battery applications. Angewandte Chemie International Edition. 2016;**55**:500-513. DOI: 10.1002/anie.201504971

[18] Zou Z, Li Y, Lu Z, Wang D, Cui Y, Guo B, et al. Mobile ions in composite solids. Chemical Reviews. 2020;**120**:4169-4221. DOI: 10.1021/acs.chemrev.9b00760

[19] Xu K. Electrolytes and interphases in Li-ion batteries and beyond. Chemical Reviews. 2014;**114**:11503-11618. DOI: 10.1021/cr500003w

[20] Deimede V, Elmasides C. Separators for Lithium-ion batteries: A review on the production processes and recent developments. Energy Technology. 2015;**3**:453-468. DOI: 10.1002/ente.201402215

[21] Gao J, Wang C, Han DW, Shin DM. Single-ion conducting polymer electrolytes as a key jigsaw piece for next-generation battery applications. Chemical Science. 2021;**12**:13248-13272. DOI: 10.1039/d1sc04023e

[22] Du Y, Yang H, Whiteley JM, Wan S, Jin Y, Lee SH, et al. Ionic covalent organic frameworks with Spiroborate linkage. Angewandte Chemie International Edition. 2016;**55**:1737-1741. DOI: 10.1002/anie.201509014

[23] Xu Q, Tao S, Jiang Q, Jiang D. Ion conduction in polyelectrolyte covalent organic frameworks. Journal of the American Chemical Society. 2018;**140**:7429-7432. DOI: 10.1021/jacs.8b03814

[24] Zhang G, Hong YL, Nishiyama Y, Bai S, Kitagawa S, Horike S. Accumulation of glassy poly(ethylene oxide) anchored in a covalent organic framework as a solid-state Li(+) electrolyte. Journal of the American Chemical Society. 2019;**141**:1227-1234. DOI: 10.1021/jacs.8b07670

[25] Liu Z, Zhang K, Huang G, Xu B, Hong YL, Wu X, et al. Highly processable covalent organic framework gel electrolyte enabled by side-chain engineering for Lithium-ion batteries. Angewandte Chemie International Edition. 2022;**61**:e202110695. DOI: 10.1002/anie.202110695

[26] Wang Y, Zhang K, Jiang X, Liu Z, Bian S, Pan Y, et al. Branched poly(ethylene glycol)-functionalized covalent organic frameworks as solid electrolytes. ACS Applied Energy Materials. 2021;**4**:11720-11725. DOI: 10.1021/acsaem.1c02426

[27] Chen H, Tu H, Hu C, Liu Y, Dong D, Sun Y, et al. Cationic covalent organic framework nanosheets for fast Li-ion conduction. Journal of the American Chemical Society. 2018;**140**:896-899. DOI: 10.1021/jacs.7b12292

[28] Li Z, Liu Z-W, Mu Z-J, Cao C, Li Z, Wang T-X, et al. Cationic covalent organic framework based all-solid-state electrolytes. Materials Chemistry Frontiers. 2020;**4**:1164-1173. DOI: 10.1039/c9qm00781d

[29] Li Z, Liu ZW, Li Z, Wang TX, Zhao F, Ding X, et al. Defective 2D covalent organic frameworks for

postfunctionalization. Advanced Functional Materials. 2020;**30**:1909267. DOI: 10.1002/adfm.201909267

[30] Jeong K, Park S, Jung GY, Kim SH, Lee YH, Kwak SK, et al. Solvent-free, single Lithium-ion conducting covalent organic frameworks. Journal of the American Chemical Society. 2019;**141**:5880-5885. DOI: 10.1021/jacs.9b00543

[31] Li X, Hou Q, Huang W, Xu H-S, Wang X, Yu W, et al. Solution-processable covalent organic framework electrolytes for all-solid-state Li–organic batteries. ACS Energy Letters. 2020;**5**:3498-3506. DOI: 10.1021/acsenergylett.0c01889

[32] Winter M, Brodd RJ. What are batteries, fuel cells, and supercapacitors? Chemical Reviews. 2004;**104**:4245-4269. DOI: 10.1021/cr020730k

[33] Sahoo R, Mondal S, Pal SC, Mukherjee D, Das MC. Covalent-organic frameworks (COFs) as proton conductors. Advanced Energy Materials. 2021;**11**:2102300. DOI: 10.1002/aenm.202102300

[34] He G, Li Z, Zhao J, Wang S, Wu H, Guiver MD, et al. Nanostructured ion-exchange membranes for fuel cells: Recent advances and perspectives. Advanced Materials. 2015;**27**:5280-5295. DOI: 10.1002/adma.201501406

[35] Chandra S, Kundu T, Dey K, Addicoat M, Heine T, Banerjee R. Interplaying intrinsic and extrinsic proton conductivities in covalent organic frameworks. Chemistry of Materials. 2016;**28**:1489-1494. DOI: 10.1021/acs.chemmater.5b04947

[36] Peng Y, Xu G, Hu Z, Cheng Y, Chi C, Yuan D, et al. Mechanoassisted synthesis of sulfonated covalent organic frameworks with high intrinsic proton conductivity. ACS Applied Materials & Interfaces. 2016;**8**:18505-18512. DOI: 10.1021/acsami.6b06189

[37] Liu L, Yin L, Cheng D, Zhao S, Zang HY, Zhang N, et al. Surface-mediated construction of an ultrathin free-standing covalent organic framework membrane for efficient proton conduction. Angewandte Chemie International Edition. 2021;**60**:14875-14880. DOI: 10.1002/anie.202104106

[38] Xu H, Tao S, Jiang D. Proton conduction in crystalline and porous covalent organic frameworks. Nature Materials. 2016;**15**:722-726. DOI: 10.1038/nmat4611

[39] Wu X, Hong YL, Xu B, Nishiyama Y, Horike S. Perfluoroalkyl-functionalized covalent organic frameworks with Superhydrophobicity for anhydrous proton conduction. Journal of the American Chemical Society. 2020;**142**:14357-14364. DOI: 10.1021/jacs.0c06474

[40] Tao S, Zhai L, Dinga Wonanke AD, Addicoat MA, Jiang Q, Jiang D. Confining H3PO4 network in covalent organic frameworks enables proton super flow. Nature Communications. 2020;**11**:1981. DOI: 10.1038/s41467-020-15918-1

[41] Li J, Wang J, Wu Z, Tao S, Jiang D. Ultrafast and stable proton conduction in polybenzimidazole covalent organic frameworks via confinement and activation. Angewandte Chemie International Edition. 2021;**60**:12918-12923. DOI: 10.1002/anie.202101400

[42] Li P, Chen J, Tang S. Ionic liquid-impregnated covalent organic framework/silk nanofibril composite membrane for efficient proton conduction. Chemical Engineering Journal. 2021;**415**:129021. DOI: 10.1016/j.cej.2021.129021

[43] Guo Y, Zou X, Li W, Hu Y, Jin Z, Sun Z, et al. High-density sulfonic acid-grafted covalent organic frameworks with efficient anhydrous proton conduction. Journal of Materials Chemistry A. 2022;**10**:6499-6507. DOI: 10.1039/d2ta00793b

[44] Wang YJ, Qiao J, Baker R, Zhang J. Alkaline polymer electrolyte membranes for fuel cell applications. Chemical Society Reviews. 2013;**42**:5768-5787. DOI: 10.1039/c3cs60053j

[45] Chen C, Tse YL, Lindberg GE, Knight C, Voth GA. Hydroxide solvation and transport in anion exchange membranes. Journal of the American Chemical Society. 2016;**138**:991-1000. DOI: 10.1021/jacs.5b11951

[46] Li N, Guiver MD. Ion transport by nanochannels in ion-containing aromatic copolymers. Macromolecules. 2014;**47**:2175-2198. DOI: 10.1021/ma402254h

[47] Akiyama R, Yokota N, Otsuji K, Miyatake K. Structurally well-defined anion conductive aromatic copolymers: Effect of the side-chain length. Macromolecules. 2018;**51**:3394-3404. DOI: 10.1021/acs.macromol.8b00284

[48] Han J, Pan J, Chen C, Wei L, Wang Y, Pan Q, et al. Effect of micromorphology on alkaline polymer electrolyte stability. ACS Applied Materials & Interfaces. 2019;**11**:469-477. DOI: 10.1021/acsami.8b09481

[49] Tao S, Xu H, Xu Q, Hijikata Y, Jiang Q, Irle S, et al. Hydroxide anion transport in covalent organic frameworks. Journal of the American Chemical Society. 2021;**143**:8970-8975. DOI: 10.1021/jacs.1c03268

[50] He G, Zhang R, Jiang Z. Engineering Covalent Organic Framework Membranes. Accounts of Materials Research. 2021;**2**:630-643. DOI: 10.1021/accountsmr.1c00083

[51] Wang X, Shi B, Yang H, Guan J, Liang X, Fan C, et al. Assembling covalent organic framework membranes with superior ion exchange capacity. Nature Communications. 2022;**13**:1020. DOI: 10.1038/s41467-022-28643-8

[52] Sasmal HS, Halder A, Kunjattu HS, Dey K, Nadol A, Ajithkumar TG, et al. Covalent self-assembly in two dimensions: Connecting covalent organic framework nanospheres into crystalline and porous thin films. Journal of the American Chemical Society. 2019;**141**:20371-20379. DOI: 10.1021/jacs.9b10788

[53] He X, Yang Y, Wu H, He G, Xu Z, Kong Y, et al. De novo Design of Covalent Organic Framework Membranes toward ultrafast anion transport. Advanced Materials. 2020;**32**:e2001284. DOI: 10.1002/adma.202001284

[54] Kong Y, He X, Wu H, Yang Y, Cao L, Li R, et al. Tight covalent organic framework membranes for efficient anion transport via molecular precursor engineering. Angewandte Chemie International Edition. 2021;**60**:17638-17646. DOI: 10.1002/anie.202105190

[55] Zhao G, Hu L, Jiang J, Mei Z, An Q, Lv P, et al. COFs-based electrolyte accelerates the Na^+ diffusion and restrains dendrite growth in quasi-solid-state organic batteries. Nano Energy. 2022;**92**:106756. DOI: 10.1016/j.nanoen.2021.106756

[56] Park S, Kristanto I, Jung GY, Ahn D, Jeong K, Kwak SK, et al. A single-ion conducting covalent organic framework for aqueous rechargeable Zn-ion batteries. Chemical Science. 2020;**11**:11692-11698. DOI: 10.1039/D0SC02785E

Chapter 3

Photoredox Catalysis by Covalent Organic Frameworks

Shuai Bi

Abstract

In recent years, photocatalysis that uses solar energy for either fuel production, such as hydrogen evolution and hydrocarbon production, or directed organic transformations, has shown great potential to achieve the goal of finding clean and renewable energy sources. Covalent organic frameworks (COFs) are crystalline organic porous materials formed by the covalent bonding of organic building blocks, which features superior structural regularity, robust framework, inherent porosity, and diverse functionality. The introduction of organic monomers with adjustable light absorption ability into COFs can make them show strong potential in photocatalysis. This chapter presents the recent progress of COF-based photocatalysts. The use of COF photocatalysts in a myriad of photoredox catalysts with a range of applications, including photocatalytic water splitting, photocatalytic CO_2 reduction, photocatalytic organic transformations, and photocatalytic environmental pollutant degradation will be highlighted. Furthermore, various linkers between COF building blocks such as nitrogen-containing connections and all sp^2-carbon connections will be summarized and compared. Finally, a perspective on the opportunities and challenges for the future development of COF and COF-based photocatalysts will be given.

Keywords: covalent organic frameworks, photocatalytic hydrogen evolution, photocatalytic CO_2 reduction, photocatalytic organic reaction, photocatalytic degradation of pollutants

1. Introduction

Covalent organic frameworks (COFs) materials are a new class of crystalline organic porous polymers, which are formed by the expansion of organic monomers through covalent bonds in two or three dimensions [1–4]. In 2005, Yaghi et al. reported the synthesis of two kinds of crystalline organic porous materials, namely COF-1 and COF-5 for the first time through trimerization of aromatic boric acid compounds, which opens the research of COFs materials [5]. By elaborately designing the organic monomers that constitute COFs, the precise control of material structures and functions can be realized [6]. By virtue of the topology design theory in Reticular Chemistry, COFs have developed rapidly in recent years [7, 8]. So far, the linking mode of COFs has developed from the initial B−O bond to C=N bond and then to C−O bond and C=C bond, etc. [9]. In terms of spatial topology,

COFs can be divided into two-dimensional (2D) COFs and three-dimensional (3D) COFs [10]. The 2D COFs are similar to graphite in structure. The organic monomers in the layers are connected by covalent bonds to form a 2D plane, and the interlayers are stacked by π-π conjugated interaction. 3D COFs are constructed by using non-planar organic monomers that expand in 3D direction through covalent bonds. 3D COFs have various types of topological structures, usually with interpenetrated structures.

COFs materials usually have the characteristics of high structure order, large specific surface area, and good thermal stability, which endow this kind of material with a wide range of applications in gas storage and separation, catalysis, sensing, energy storage, and photoelectric conversion [11]. In recent years, with more in-depth research on COFs, it has been found that COFs can generate and separate charge carriers under the irradiation of light. Combined with their designable pore structure, COFs have been witnessed as potential photocatalysts [12–14].

Photocatalysis uses light energy to drive chemical reactions, which is a green catalytic method and has great research value. In recent years, a large number of photocatalysts suitable for different systems have been developed [15]. Generally, photocatalysts are divided into homogeneous and heterogeneous photocatalysts. Compared with homogeneous photocatalysts, heterogeneous photocatalysts have the advantages of recyclability and reusability while maintaining high catalytic activity, which makes heterogeneous photocatalysts more potential for industrial application. However, current heterogeneous photocatalysts based on inorganic metal oxides and other systems still face some challenges, such as weak light absorption ability and poor charge separation ability [16]. In this context, recently developed COFs materials have unique advantages in the field of photocatalysis: (1) Structure designability: The light absorption capacity and energy band structure of COFs can be finely adjusted by changing the chemical structure of the monomers. (2) Facile functionalization. The COFs skeleton can be modified in two ways, before and after preparation, so that the catalytic sites can be easily introduced into the COF skeleton. (3) High surface area and good stability. This allows the catalytic sites on the COFs to be effectively dispersed and fully exposed and to remain stable during the catalytic process. In this chapter, the research progress of COFs in photocatalytic hydrogen production, photocatalytic carbon dioxide reduction, photocatalytic organic reactions, and photocatalytic pollutant degradation is summarized, and the future application of COFs in the field of photocatalysis is also prospected.

2. Photocatalytic hydrogen production

Hydrogen energy is recognized as clean energy. Photocatalytic hydrogen production from water splitting by using solar energy is a very promising approach. At present, the main obstacle to photocatalytic hydrogen production is the lack of efficient catalysts. The structural diversity of COFs enables them excellent visible light absorption ability, tunable band position, and good stability. These advantages make COFs expected to become efficient catalysts in the field of photocatalytic hydrogen production. In recent years, COFs as catalysts for photocatalytic hydrogen production have gradually attracted researchers' attention. Up to now, there are mainly three kinds of COFs photocatalytic hydrogen production systems: pure COFs photocatalysts, COFs-based heterojunction photocatalysts, and COFs composites photocatalysts.

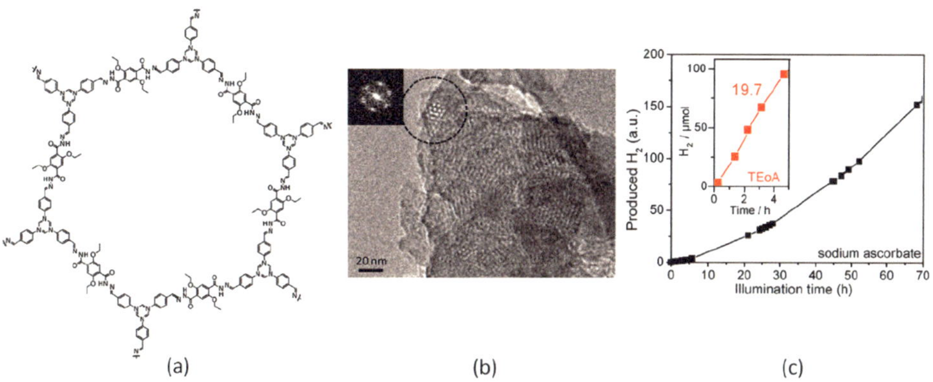

Figure 1.
(a) Chemical structure of TFPT-COF. (b) High-resolution transmission electron microscope (TEM) images. (c) Photocatalytic H_2 evolution with TFPT-COF using sodium ascorbate donor and Pt cocatalyst under visible light irradiation. Inset: H_2 evolution using TEOA as an electron donor. Adapted with permission from Ref. [17]. Copyright 2014 Royal Society of Chemistry.

2.1 Pure COFs photocatalysts

In 2014, Lotsch et al. used 1,3,5-tris(4-formyl-phenyl)triazine (TFPT) and 2,5-diethoxy-terephthalohydrazide as building blocks to construct hydrazone-linked TFPT-COF and used this COF to realize the photocatalytic hydrogen production from proton reduction in water (**Figure 1**) [17]. Under the irradiation of visible light, in the presence of metallic platinum that was used as the cocatalyst, and triethanolamine used as the hole sacrificial agent, the hydrogen production rate of this COF reached 1970 $\mu mol \cdot h^{-1} \cdot g^{-1}$ with an apparent quantum efficiency of 2.2%. The authors speculated that the central triazine unit with good planarity in this COF enables enhanced stacking interactions and thus charge transport in the axial direction, rendering TFPT-COF an effective H_2-evolving photocatalytic system.

Lotsch et al. further explored the effect of the number of nitrogen atoms in the benzene ring at the central node of COF on the photocatalytic activity [18]. A series of stable azine-linked N_x-COFs, named N_0-COF, N_1-COF, N_2-COF, and N_3-COF, were synthesized by condensation of hydrazine and triphenylaryl monomers with 0, 1, 2, or 3 nitrogen atoms in the central benzene ring, respectively. Under visible light irradiation, with platinum as the cocatalyst and triethanolamine as the hole sacrificial agent, the hydrogen production rates of this kind of COF gradually increased with the increase of nitrogen content, and the hydrogen production rates from N_0-COF to N_3-COF were 23, 90, 438, and 1703 $\mu mol \cdot h^{-1} \cdot g^{-1}$, respectively. This indicates that N_3-COF has a high photocatalytic activity, which may be attributed to the stabilization of photogenerated radical anions by the electron-deficient triazine center, which enhances the charge separation ability and facilitates the electron migration to the platinum cocatalyst.

In 2017, Lotsch et al. further demonstrated that both the number and position of nitrogen atoms had a great impact on the photocatalytic activity of COFs [19]. PTP-COF has a similar structure to N_3-COF with three nitrogen atoms located on the pyridine group at the periphery of the 1,3,5-tripyridyl benzene monomer. Under the same photocatalytic hydrogen production conditions as N_3-COF, the hydrogen production rate of PTP-COF is only 83.83 $\mu mol \cdot h^{-1} \cdot g^{-1}$. The authors further studied the effect of the number of nitrogen atoms in the peripheral aromatic ring of COF monomer on the photocatalytic hydrogen production activity. Three azine-linked COFs were synthesized by condensation of hydrazine with three monomers containing 0, 1, or 2 nitrogen atoms in the peripheral aryl group of 1,3,6,8-tetrakis(4-ethynylbenzaldehyde)-pyrene [20]. The hydrogen production rates of these COFs decreased gradually with the increase of nitrogen content, which were 98, 22, and 6 μmol·h^{-1}·g^{-1}, respectively. It is obvious that the subtle structural changes of monomers have significant effects on photocatalytic hydrogen production of COFs, but the underlying mechanism needs to be further confirmed by theory and experiment.

The photocatalytic hydrogen production activity of COFs can be changed by designing different monomer structures. Therefore, in recent years, the diacetylene- and thiophene-based COF photocatalysts systems have been developed. Thomas et al. synthesized TP-BDDA-COF and TP-EDDA-COF containing acetylene fragments [21]. The diacetylene groups in TP-BDDA-COF possess larger conjugation, which endows TP-BDDA-COF with a narrower band gap and faster charge mobility. Under visible light irradiation, with platinum as cocatalyst and triethanolamine as hole sacrificial agent, the hydrogen production rate of TP-BDDA-COF is 324 μmol·h^{-1}·g^{-1}, which is about 10 times that of TP-EDDA-COF that is linked by only acetylene group (30 μmol·h^{-1}·g^{-1}). Cooper et al. constructed S-COF and FS-COF by using 3,7-diaminodibenzo[*b*,*d*]thiophene and 3,9-diamino-benzo[1,2-*b*:4,5-*b*']bis[1]benzothiophene sulfone as monomers, respectively [22]. The key dibenzo[*b*,*d*]thiophene sulfone (DBTS) unit is the monomer of photosensitive polymer P10 [23]. Under visible light irradiation, with platinum as cocatalyst and sodium ascorbate as hole sacrificial agent, the hydrogen production rate of FS-COF was 10.1 mmol·h^{-1}·g^{-1}, while that of S-COF was 4.44 mmol·h^{-1}·g^{-1}. Without platinum as a cocatalyst, FS-COF and S-COF could still produce hydrogen efficiently with hydrogen production rates of 1.32 and 0.6 mmol·h^{-1}·g^{-1}, respectively. They claim that the higher performance of FS-COF is due to the characteristics, such as larger pore size, higher specific surface area, wider light absorption range, and more hydrophilic pore structure, which make FS-COF have better charge transfer ability and more abundant photocatalytic sites. They further investigated the effect of crystallinity of COF on photocatalysis, showing that crystalline FS-COF had higher photocatalytic activity compared with amorphous or semi-crystalline FS-P. They also improved the photocatalytic activity of FS-COF by organic dyes sensitization, enhancing the hydrogen production rate to 16.3 mmol·h^{-1}·g^{-1}.

The functional groups on the skeleton of COFs have a strong influence on their photocatalytic activity. Sun et al. obtained TpPa-COF, TpPa-COF-$(CH_3)_2$, and TpPa-COF-NO_2 by condensation of *p*-phenylenediamine, 2,5-dimethyl-p-phenylenediamine, 2-nitro-*p*-phenylenediamine, and 1,3,5-trimethylphloroglucinol, respectively [24]. Under visible light irradiation, with platinum as cocatalyst and sodium ascorbate as hole sacrificial agent, the hydrogen production rates were 1.56, 8.33, 0.22 mmol·h^{-1}·g^{-1}, respectively. Through a series of characterizations, they proved that the electron donor groups in COFs have a strong conjugation effect, which enhanced the light absorption and carrier mobility of the material, thus improving the photocatalytic activity of the material.

In 2019, Zhang et al. reported a series of vinylene-linked (−HC=CH−) 2D COFs through Knoevenagel condensation of 3,5-dicyano-2,4,6-trimethylpyridine (DCTMP) with different aromatic aldehyde monomers (**Figure 2**) [25]. The polymerization and crystallization processes were performed in *N*,*N*-dimethylformamide (DMF) with piperidine as a catalyst. Upon reaction of DCTMP with 4,4-diformyl-*p*-terphenyl (DFPTP), 4,4-diformyl-1,1-biphenyl (DFBP), and 1,3,5-tri(4-formylphenyl)

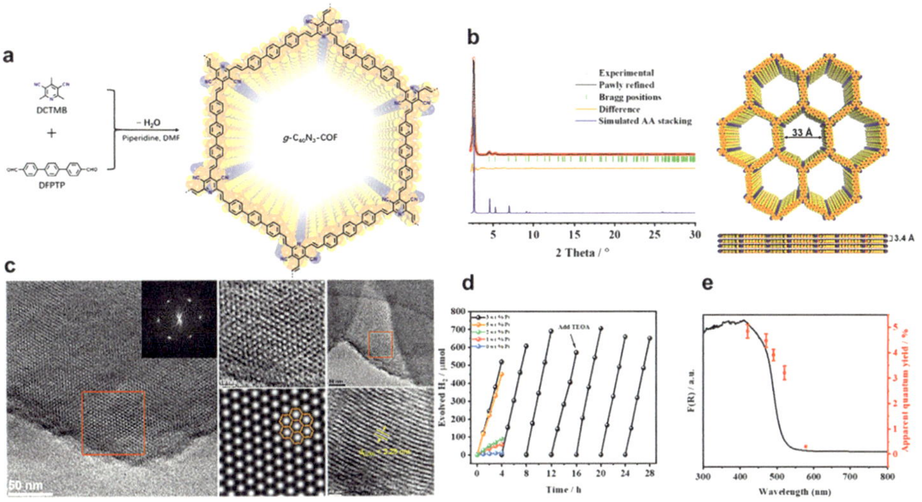

Figure 2.
(a) Synthesis of the vinylene-linked g-$C_{40}N_3$-COF. (b) Powder X-ray diffraction (PXRD) analysis and structural simulation of g-$C_{40}N_3$-COF. (c) High-resolution TEM images of g-$C_{40}N_3$-COF. (d) Photocatalytic H_2 evolution performance of Pt@g-$C_{40}N_3$-COF. (e) Apparent quantum yield (AQY) of g-$C_{40}N_3$-COF along with its UV-vis diffuse reflectance spectra (DRS). Adapted with permission from Ref. [25]. Copyright 2019, Springer Nature.

benzene (TFPB) respectively, three 2D COFs (denoted as g-C_xN_y-COF) with *trans*-disubstituted C=C bond linkages (vinylene linkages) are formed in high yields. The fully aromatic structures of the frameworks containing pyridine rings, benzene rings, vinylene bonds, and cyano groups grant them intrinsic π-electron delocalized properties and excellent light-harvesting capabilities. In addition, the COFs showed ordered structures with high crystallinity, high surface area up to 1235 $m^2\ g^{-1}$, and appropriate band structures, which allowed them to drive two half-reactions of water splitting to generate hydrogen or oxygen separately under visible light irradiation. The sample g-$C_{40}N_3$-COF exhibited the most prominent performance with a high H_2 production rate of 4120 $\mu mol{\cdot}h^{-1}{\cdot}g^{-1}$ and a remarkable apparent quantum efficiency (AQY) of 4.84% (at a wavelength of 420 nm). These values exceeded all the other COF-based photocatalysts at that time, demonstrating the superior catalytic activities of this kind of COF.

In the same year, by using another key monomer 2,4,6-trimethyl-1,3,5-triazine (TMTA) to react with aromatic multi-aldehydes, Zhang et al. established two triazine-cored COFs with unsubstituted carbon-carbon double bond linkages [26]. The resulting sp^2-carbon-linked 2D COFs showed high crystallinity, extended π-electron delocalization, tunable band energy levels, as well as high surface areas, regular open channels, and superior chemical stabilities. These features endow the COFs with potential as semiconducting photocatalysts. Moreover, these COFs showed fibrous morphologies, which allowed for fabrication of thin films as photoelectrodes by a simple drop-casting performance. The corresponding photoelectrodes exhibited photocurrents up to 45 $\mu A\ cm^{-2}$ at 0.2 V vs. RHE, which is significantly higher than that of its imine-linked COF analogue, demonstrating the superior semiconducting properties of the vinylene-linked COFs. In the particulate photocatalysis system, a 3% Pt-modified sample exhibited an H_2 production rate of 292 $\mu mol{\cdot}h^{-1}{\cdot}g^{-1}$ with an AQY of 1.06% at 420 nm upon visible light irradiation, outperforming its imine-linked

COF analogue that nearly lost its crystallinity upon 4-h light exposure, indicating the superior photostability of vinylene-linked COFs over their imine-linked COF counterparts.

In a subsequent work, Zhang et al. developed two semiconducting vinylene-linked COFs with aligned octupolar structures and tunable polarity. The polarity of the COFs was tuned by reticulating D_{3h}-symmetric 2,4,6-tris(4′-formyl-biphenyl-4-yl)-1,3,5-triazine with D_{3h}-symmetric tricyanomesitylene (g-$C_{54}N_6$-COF) and C_{2v}-symmetric 3,5-dicyano-2,4,6-trimethylpyridine (g-$C_{52}N_6$-COF), respectively [27]. The highly symmetric g-$C_{54}N_6$-COF exhibited the octupolar conjugated characters with more promising semiconducting behavior as compared with the less-symmetric g-$C_{52}N_6$-COF. The octupolar π-conjugated g-$C_{54}N_6$-COF exhibited enhanced light harvesting as well as excellent photo-induced charge generation and separation capabilities. Due to the appropriate band energy levels that are suitable for water splitting, g-$C_{54}N_6$-COF enabled the two half-reactions of photocatalytic water splitting with an average O_2 production rate of 51.0 $\mu mol \cdot h^{-1} \cdot g^{-1}$ and H_2 production rate of 2518.9 $\mu mol \cdot h^{-1} \cdot g^{-1}$, indicating the immense prospect of vinylene-linked COFs for photocatalytic overall water splitting. In summary, g-$C_{54}N_6$-COF showed significantly higher photocatalytic performance than the less-symmetric g-$C_{52}N_6$-COF, implying that changing the geometric symmetry of monomers within vinylene-linked COFs could exert profound effects on their polarity, semiconducting properties, as well as photocatalytic performance.

The robust solid-state crystalline structure of COFs is fully modular and hence possesses almost unlimited chemical tunability for the different functions fundamental to the photocatalytic process at an atomic level precision. Among the reported COF-based photocatalysts, vinylene-linked COFs showed higher H_2 evolution efficiency than the other COFs, probably due to their superior π-electron conjugation, excellent light-harvesting capability, fast charge isolation, high carrier mobility, and exceptional stability. However, a wide range of structural and optoelectronic factors need to be well orchestrated to maximize the H_2 evolution efficiency of a COF photocatalyst. Engineering the chemistry of COF—sacrificial electron donor and the COF—cocatalyst interfaces, or creating COF-based heterojunction composites would be vital for further improvement of H_2 evolution efficiency.

2.2 COFs-based heterojunction photocatalysts

To further enhance the separation of photogenerated charges, researchers designed COFs heterojunction catalysts, which were formed by covalent bonding. There are two main design strategies for this type of catalyst: (1) introducing donor and acceptor units into the COFs to form heterojunction catalyst; (2) composite COFs with other materials to form heterojunction catalyst. These two strategies can effectively tune the energy band structure of the photocatalyst and increase the separation efficiency of photogenerated electrons and holes.

Jiang et al. reported a newly developed photocatalyst that consists of all sp^2 carbon frameworks with full π-conjugation (**Figure 3**) [28]. The COF was synthesized by Knoevenagel condensation between tetrakis(4-formylphenyl)pyrene and 1,4-phenylenediacetonitrile (PDAN). Further, the framework was designed to constitute built-in donor-acceptor heterojunction by integrating the electron deficient 3-ethylrhodanine (ERDN) unit as an end-capping group to the periphery of the sp^2c-COF lattice. The resultant COF created dense yet ordered columnar π-arrays, electron push-and-pull effect, and substantial reaction centers in pores or on surface, which could promote exciton migration and offer a narrow band gap to harvest visible and

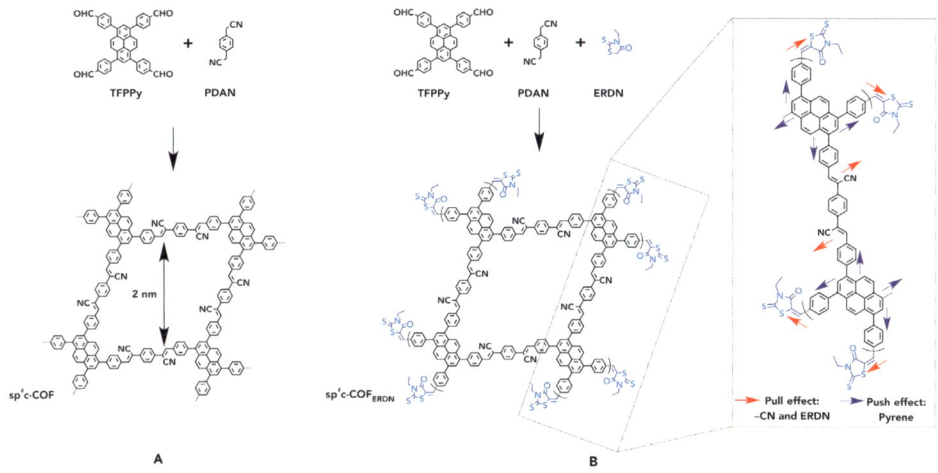

Figure 3.
Design and synthesis of the 2D stable sp² carbon-conjugated COFs for H_2 production from water. Schematic representation of (A) sp²c-COF, (B) sp²c-COFERDN with electron-deficient ERDN end groups. Inset: electron donor-acceptor pull-push effects on the 2D skeletons. Adapted with permission from Ref. [28]. Copyright 2019, Elsevier Inc.

near-infrared light. Consequently, the COFs enabled an immediate yet continuous stable hydrogen production from water upon irradiation. These results demonstrated that the sp^2 carbon frameworks ensemble a light-driven photocatalytic system in which a chain of photochemical events—including light harvesting, exciton migration, exciton splitting, and electron transfer and collection—is structurally interlocked and seamlessly coupled. One could rationally design the COF photocatalysts with specific semiconducting properties for the continuous and efficient production of hydrogen from water.

Heterojunctions between COFs and other materials can be formed by covalently bonding COFs with semiconductors that have different energy band structures. For example, COF and g-C_3N_4 are connected by the imine bond to form heterojunction CN-COF. CN-COF can produce hydrogen under visible light with high photocatalytic activity [29]. The rate can reach 10.1 $mmol \cdot h^{-1} \cdot g^{-1}$, which is mainly due to the suitable light absorption range and reasonable energy band structure of the heterojunction material. In addition, covalent bonding can change the photogenerated electron transport path, inhibit carrier recombination, and effectively improve photocatalytic activity.

The formation of heterojunction materials between metal-organic frameworks (MOFs) and COFs is a new way to improve photocatalytic activity. NH_2-UiO-66 was immobilized on the surface of TpPa-1-COF by the covalent bond, and the maximum hydrogen yield rate was 23.41 $mmol \cdot h^{-1} \cdot g^{-1}$ under visible light [30]. In this photocatalytic process, the photogenerated electrons of TpPa-1-COF transition from the valence band to the conduction band and are rapidly transferred to the conduction band of NH_2-UiO-66 by heterojunction, so as to ensure the effective separation of photogenerated electrons and holes, and thus improve the catalytic efficiency.

2.3 COFs composites photocatalysts

In addition to the advantages of COFs, such as facile structural designability, compositing COFs with other materials to form photocatalysts is also an important strategy that is developed in recent years. COFs that composite with other materials can not only make full use of their own advantages but also greatly improve the separation and transport efficiency of photogenerated carriers.

Banerjee et al. synthesized β-ketoenamine-linked COF (TpPa-2 COF) with extended π-conjugation, high specific surface area, and excellent stability [31]. Subsequently, CdS nanoparticles were uniformly dispersed in the synthesized TpPa-2 COF to form a composite photocatalyst. Two-dimensional TpPa-2 COF can not only stabilize CdS nanoparticles within its pore channels but also promote charge transfer and inhibit photogenerated carrier recombination, thus improving photocatalytic activity. Under visible light irradiation, the hydrogen production rate of the CdS-COF composite catalyst was up to 3678 $\mu mol{\cdot}h^{-1}{\cdot}g^{-1}$ with platinum as cocatalyst and lactic acid as hole sacrificial agent. However, the hydrogen production rate of CdS nanoparticles was 128 $\mu mol{\cdot}h^{-1}{\cdot}g^{-1}$, and that of COF alone was only 28 $\mu mol{\cdot}h^{-1}{\cdot}g^{-1}$.

COFs combined with other noble metal nanoparticles can also effectively inhibit the recombination of photogenerated carriers and improve photocatalytic ability. Long et al. coated Pt nanoparticles with polyvinylpyrrolidone (PVP) and assembled the coated Pt nanoparticles with hydrophilic imine-linked TP-COF by electrostatic assembly method to prepare a metal-insulator-semiconductor (MIS) photocatalytic system [32]. Under visible light irradiation, the hydrogen production rate of the ternary PT-PVP-TP-COF photocatalyst is 8.42 $mmol{\cdot}h^{-1}{\cdot}g^{-1}$, which is 32 times higher than that of the corresponding metal-semiconductor Mott-Schottky photocatalyst. The authors clarify that the ternary system can extract photogenerated electrons from COF with the help of an electrostatic field and transfer them to platinum nanoparticles through an insulator. However, the existence of an insulator prevents the photogenerated holes generated in COF from migrating to platinum nanoparticles, thus effectively improving the separation efficiency of electrons and holes and suppressing the recombination of photogenerated carriers.

In the process of photocatalytic hydrogen production, platinum is usually needed as a cocatalyst to facilitate the proton reduction to hydrogen in the system. However, the use of platinum increases the cost and hinders the industrial application of photolytic water splitting. In order to solve this problem, Lotsch et al. developed a photocatalytic hydrogen production system with chloro(pyridine)cobaloxime as cocatalyst (**Figure 4**) [33]. With triethanolamine as the hole sacrificial agent, the hydrogen production rate of N2-COF can reach 782 $\mu mol{\cdot}h^{-1}{\cdot}g^{-1}$. This work represents an economic system for photocatalytic H_2 evolution using COF photosensitizers and earth-abundant cocatalysts. Recently, Lotsch et al. used an earth-abundant, noble-metal-free nickel-thiolate hexameric cluster as a cocatalyst for photocatalytic hydrogen production [34]. Under the condition of visible light irradiation and triethanolamine as hole sacrificial agent, TpDTz COF synthesized from thiazolo[5,4-*d*] thiazole monomer was used as a photocatalyst, delivering the hydrogen production rate of 941 $\mu mol{\cdot}h^{-1}{\cdot}g^{-1}$.

3. Photocatalytic CO_2 reduction

With the development of human society, the CO_2 content in the atmosphere is increasing rapidly, which leads to the greenhouse effect. It is urgent for human beings to find a green and convenient way to convert CO_2 into usable chemicals. The CO_2 molecule has a highly symmetric structure and is chemically relatively inert. Although chemical and electrochemical methods have demonstrated excellent CO_2 conversion

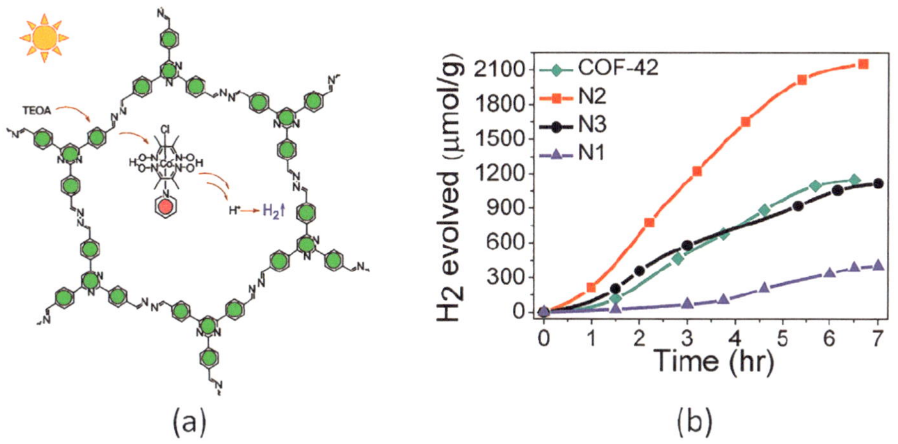

Figure 4.
(a) Schematic illustration of photocatalytic H_2 evolution by using N_2-COF photosensitizer and Co-1 cocatalyst. (b) H_2 evolution rates with various COF photosensitizers using Co-1 cocatalyst and TEOA donor. Adapted with permission from Ref. [33]. Copyright 2017, American Chemical Society.

capabilities, the direct use of light energy to achieve CO_2 conversion, especially CO_2 reduction, still presents many challenges. Photocatalytic CO_2 reduction requires not only high stability and strong visible light absorption capacity but also strong CO_2 adsorption capacity and suitable photocatalytic sites. On the basis of high crystallinity, high specific surface area, and adjustable light absorption ability, the introduction of suitable CO_2 catalytic sites will make COFs become potential CO_2 heterogeneous photocatalysts.

3.1 Introduction of metal sites into COFs for photocatalytic CO_2 reduction

Rhenium complexes are excellent homogeneous CO_2 reduction photocatalysts. Cao et al. synthesized pyridine-containing covalent triazene framework (CTF) materials and used pyridine nitrogen atoms in the organic skeleton to anchor $Re(CO)_3Cl$ complexes [35]. The obtained photocatalyst Re-CTF-py with CO_2 reduction ability has high stability, strong CO_2 adsorption ability, and good photoactivity. Under visible light irradiation and triethanolamine as a hole sacrificial agent, the rate of CO_2 reduction to CO is 353. 05 $\mu mol{\cdot}h^{-1}{\cdot}g^{-1}$ with high selectivity. Due to the coordination of Re and pyridine in Re-CTF-py, the leaching of $Re(CO)_3Cl$ can be effectively avoided, exhibiting excellent recyclability. Similarly, Huang et al. synthesized imine-linked COF using 2,2′-bipyridyl-5,5′-dialdehyde (BPDA) and 4,4′,4″-(1,3,5-triazine-2,4,6-triyl)trianiline (TTA), which anchored $Re(CO)_5Cl$ at the adjacent nitrogen atoms in bipyridine units to form COF-based photocatalyst (**Figure 5**) [36]. The COF can reduce CO_2 to CO with a selectivity of up to 98% and a reaction rate of 750 $\mu mol{\cdot}h^{-1}{\cdot}g^{-1}$. The reduction mechanism was investigated by transient spectroscopy, X-ray absorption spectroscopy, and in situ diffuse reflectance spectroscopy. It was concluded that photogenerated electrons could be rapidly transferred from COF to Re catalytic sites. In recent work, TpBpy COF chelates $Ni(ClO_4)_2$ through the bipyridine units to form Ni-TpBpy COF containing a single atomic Ni site [37]. With $Ru(bpy)_3Cl_2$ as a photosensitizer, triethanolamine as a hole sacrificial agent, and Ni-TpBpy COF as cocatalyst, the photocatalytic reduction rate of CO_2 to CO can reach 811.4 $\mu mol{\cdot}h^{-1}{\cdot}g^{-1}$ with the selectivity of 96%. In the Ni-TpBpy COF system, COF can not only serve as a

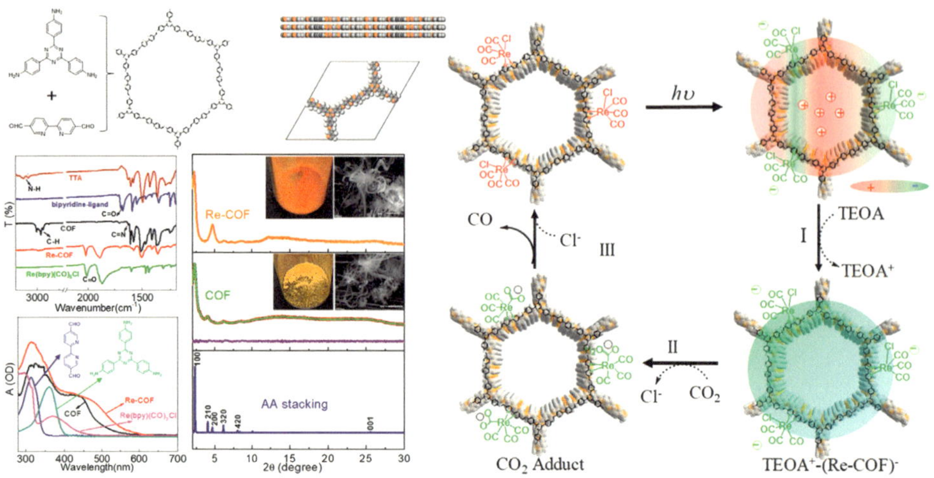

Figure 5.
Synthesis and characterization of COF and Re-COF. Proposed catalytic mechanism for CO_2 reduction. Adapted with permission from Ref. [36]. Copyright 2018, American Chemical Society.

carrier of the CO2 catalytic site but also promote the stability of the intermediate and improve the reduction efficiency of CO_2.

Oxygen atoms in COFs can also act as metal anchoring sites. For example, Lan et al. synthesized DQTP COF with 2,6-diaminoanthraquinone as the monomer, in which the oxygen atoms of quinone between the adjacent layers have suitable distances, which provides a feasible coordination environment for anchoring transition metal catalysts [38]. Interestingly, CO_2 can be selectively reduced to different products after introducing different metal catalytic sites. DQTP COF coordinates with Zn(II), Co(II), and Ni(II) to form DQTP COF-M photocatalyst, respectively. CO_2 was reduced by using $Ru(bpy)_3Cl_2$ as a photosensitizer and triethanolamine as a hole sacrificial agent under visible light. The reduction product of DQTP COF-Co was CO at a production rate of 1020 $\mu mol{\cdot}h^{-1}{\cdot}g^{-1}$. When the coordinated metal ion is changed to Zn, the reduction product of DQTP COF-Zn is HCO_2H with a rate of 152.5 $\mu mol{\cdot}h^{-1}{\cdot}g^{-1}$, and the selectivity is greater than 90%. The authors concluded that the electron-rich coordination environment tends to form CO in the reduction process, while the electron-deficient coordination environment tends to form HCO_2H. CO(II), as a good π-bond donor, is conducive to the conversion of CO_2 to CO. However, Zn(II) is a poor donor of the π-bond, so its reaction product is mainly HCO_2H. The photocatalytic CO_2 reduction products of DQTP COF-Ni with a moderate electron coordination environment are almost equal amounts of CO and HCO_2H.

Metal porphyrins are excellent catalytic sites for CO_2 reduction. Introducing metal porphyrins into the COF skeleton is also an important strategy for the design of CO_2 photoreduction catalysts. Lan et al. synthesized TTCOF-Zn from electron-deficient Zn metal porphyrin and electron-rich tetrathiafulvalene [39]. In TTCOF-Zn, the donor-acceptor structure between tetrathiafulvalene and porphyrin was formed, which could effectively improve the separation ability of photogenerated charge. Under visible light irradiation, without adding any sacrificial agent or photosensitizer, TTCOF-Zn can reduce CO_2 to CO in an aqueous solution with a yield of 0. 2055 $\mu mol{\cdot}h^{-1}{\cdot}g^{-1}$, while the selectivity reaches 100%. The catalyst can still maintain good catalytic activity after five cycles.

The morphology of the catalyst also has a significant influence on the performance of the catalyst. The bulk catalyst can be nano-sized to expose more catalytic sites and thus improve catalytic activity [40]. Jiang et al. synthesized COF-367-Co nanosheets from 5,10,15,20-tetra(*p*-aminophenyl)porphyrin (H_2TAPP) and 4,4′-biphenyldialdehyde (BPDA) as monomers [41]. 2,4,6-Trimethylbenzaldehyde (TBA) can be introduced to the edge of COF-367-Co nanosheets by imine exchange reaction. TBA with large steric hindrance obstructs the axial packing of COF 2D lamellae and promotes the growth of COF nanosheets along the plane direction, resulting in ultrathin 2D COF nanosheets. Ultrathin 2D COF nanosheets can effectively expose the catalytic site of cobalt metalloporphyrin and shorten the distance of photogenerated charge carriers migrating to the catalytic site. Under visible light irradiation, with $Ru(bpy)_3Cl_2$ as a photosensitizer and ascorbic acid as a hole sacrificial agent, COF-367-Co ultrathin 2D nanosheet can reduce CO_2 to CO at a rate of 10,162 $\mu mol{\cdot}h^{-1}{\cdot}g^{-1}$ with the selectivity of 78%.

3.2 Introduction of metal-free sites into COFs for photocatalytic CO_2 reduction

In addition to using metal as the catalytic site for CO_2 reduction, nitrogen atoms in the COFs skeleton can be directly used as the catalytic site for CO_2 reduction, which can achieve a totally metal-free heterogeneous photocatalytic system. Zhu et al. synthesized ACOF-1 and N_3-COF by condensation of 1,3,5-triformylbenzene (TFB) and 2,4,6-tris(4-formylphenyl)-1,3,5-triazine (TFPT) with hydrazine, respectively [42]. Through control experiments, it was found that the triazine units in COFs not only improved the light absorption capacity but also contributed to the adsorption of CO_2 and served as catalytic sites. The electron-deficient triazine in N_3-COF can effectively stabilize the negatively charged intermediate in the catalytic process. Without any sacrificial agent, N_3-COF can reduce CO_2 to methanol in an aqueous solution under visible light irradiation at a rate of 0.57 $\mu mol{\cdot}h^{-1}{\cdot}g^{-1}$. The catalytic activity of N_3-COF is higher than that of ACOF-1 without the triazine ring.

The COF heterojunction based on the donor-acceptor structure can further improve the catalytic activity. To this end, Kong et al. designed an intramolecular donor-acceptor heterojunction CT-COF containing an electron-rich carbazole group as the donor and an electron-deficient triazine group as the acceptor [43]. CT-COF has a wide optical absorption range and narrow band gap due to the push-pull electron effect on the donor-acceptor structure, which effectively improves the separation and transportation efficiency of photogenerated carriers. They proved that the nitrogen atom in the triazine group was the catalytic site of CO_2 reduction by density functional calculation. Under visible light irradiation, the rate of CO_2 reduction to CO in an aqueous solution is 102.7 $\mu mol{\cdot}h^{-1}{\cdot}g^{-1}$ catalyzed by CT-COF without any co-catalyst or sacrificial agent.

The heterojunction structure formed by COF and inorganic semiconductors can also improve photocatalytic activity effectively. Lan et al. synthesized COF-318 from 2,3,6,7,10,11-hexahydroxytriphenylene (HHTP) and 2,3,5,6-tetrafluoro-4-pyridinecarbonitrile (TFPC) as precursors, which then formed a *Z*-scheme heterojunction catalyst by covalently bonding with inorganic semiconductor TiO_2, Bi_2WO_6, and α-Fe_2O_3, respectively (**Figure 6**) [44]. Both pyridine nitrogen and nitrile nitrogen in monomers can be used as CO_2 catalytic sites. Meanwhile, the *Z*-scheme heterojunction formed by COF and inorganic semiconductors can effectively improve

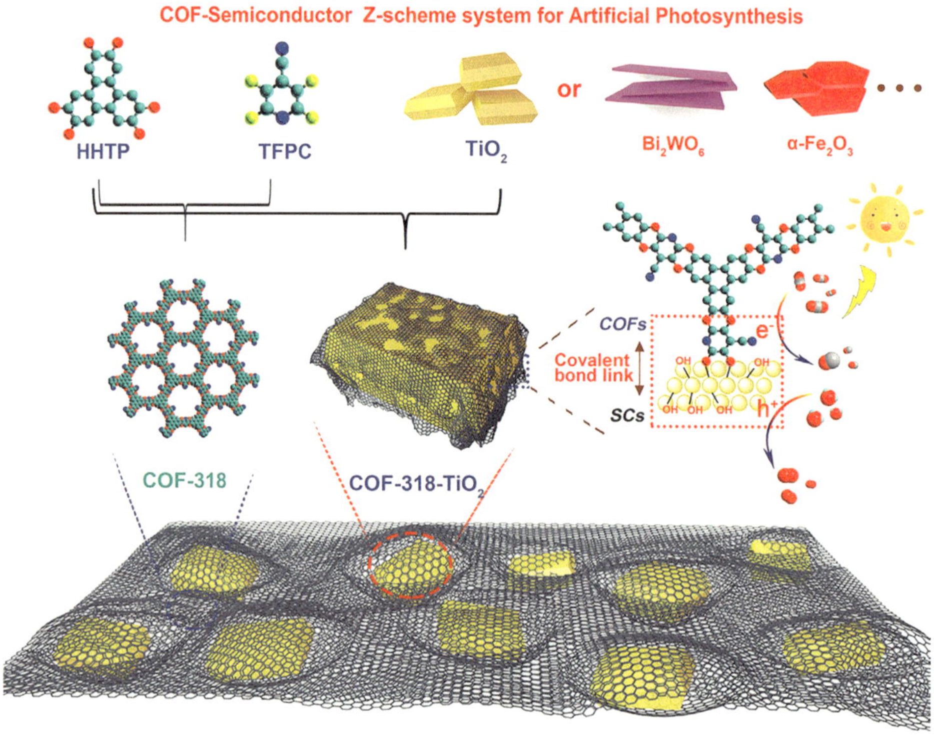

Figure 6.
Schematic representation of the preparation of COF-318-SCs via the condensation of COF-318 and semiconductor materials. Adapted with permission from Ref. [44]. Copyright 2020, Wiley-VCH.

the separation efficiency of photogenerated electrons and holes and inhibit carrier recombination. Under visible light irradiation, the reduction rate of CO_2 to CO in an aqueous solution can reach 69.67 $\mu mol \cdot h^{-1} \cdot g^{-1}$ without adding other cocatalysts.

Although the use of COFs photocatalysts to reduce CO_2 has been widely reported, the quantum efficiency of COF-based photocatalysis is still difficult to compare with other systems, and it is difficult to obtain more valuable hydrocarbon products such as ethane or ethylene through COFs photocatalysis. However, it still showed great promise, and one could rationally expect that there would be more and more research focusing on the development of more efficient and promising COF-based photocatalysts for photocatalytic CO_2 reduction.

4. Photocatalytic organic reactions

Photocatalytic organic reactions can achieve the green synthesis of organic intermediates and can obtain compounds that cannot be obtained by other methods, which plays an important role in fine organic synthesis. In recent years, a series of homogeneous photocatalysts, such as ruthenium complexes and small-molecule dyes, have greatly improved organic photocatalytic reactions in terms of both reaction types and reaction efficiency. However, the high cost and high toxicity of a series of noble metal catalysts represented by ruthenium complexes limit their application in the pharmaceutical industry. Therefore, it is particularly important to develop nontoxic and recyclable heterogeneous photocatalysts. Among them, porous materials, especially COFs, are used as organic photocatalysts, which have attracted extensive attention from researchers.

Cross-dehydrogenating coupling (CDC) is a highly efficient C—C bond coupling reaction, which can be used to synthesize a variety of important organic intermediates. For example, Liu et al. designed and synthesized COF-JLU5, which contains an electron-deficient triazine group and electron-rich 2,5-dimethoxyphenyl group to form a donor-acceptor structure [45]. The COF has the characteristics of large crystallinity, high specific surface area, high stability, and photoredox activity. The cross-coupling reaction of *N*-aryl-tetrahydroisoquinoline can be catalyzed under blue LED irradiation, with a yield up to 99%, a wide substrate range, and excellent recyclability.

Liu et al. further synthesized COF-JLU22 based on electron-rich 1,3,6,8-tetrakis (4-aminophenyl)pyrene and electron-deficient 4,4′-(benzothiadiazole-4,7-diyl) dibenzaldehyde [46]. The pyrene segment can not only enhance the crystallization of COF-JLU22 but also connect with the acceptor structure of the electron-deficient benzothiadiazole group. The resulting COF-JLU22 has a wide optical absorption range, and its band structure is very close to that of $Ru(bpy)_3Cl_2$ complexes that are widely used in photocatalysis. Conclusively, COF-JLU22 exhibited excellent catalytic performance in the reductive dehalogenation of phenacyl bromide derivatives and α-alkylation of aldehydes under visible-light irradiation and retained its original catalytic activity after three cycles.

Wu et al. synthesized a hydrazone-linked COF with high specific surface area and high crystallization, which could catalyze the CDC reaction of tetrahydroisoquinoline with nitromethane, acetone, acetophenone, and other nucleophiles under visible light, with the yield up to 87% [47]. The COF has high stability, and its catalytic activity is not reduced after three cycles.

Although imine bond-linked COFs photocatalysts show excellent catalytic performance in the field of photocatalysis, such kinds of COFs are unstable in acidic systems due to the reversibility of imine bonds. Therefore, it is very important to design and synthesize COFs with high stability for photocatalytic organic conversion.

Banerjee et al. designed and synthesized β-ketoenamine-linked COF, which has high stability [48]. TpTt COF was synthesized from 1,3,5-triazine-2,4,6-triamine and 2,4,6-triformylphloroglucinol. This COF is isomerized into the β-ketoenamine-based structure, which makes the bond of COF irreversible and thus has high chemical stability. The COF can initiate a *cis-trans* isomerization reaction of olefin under blue LED light irradiation. Due to the stability of the β-ketoenamine linkage fragment, the COF could still maintain the original photocatalytic activity after four cycles of catalytic reactions.

COFs connected by carbon-carbon double bonds possess higher chemical stability and full π-electron conjugation, representing a kind of exceptional photocatalysts. Wang et al. synthesized 2D porphyrin-based sp^2-carbon conjugated COF [49], which could maintain stability in 9 mol l^{-1} HCl and 9 mol l^{-1} NaOH. Due to the unique photophysical and redox properties of porphyrin, as well as the high stability and π-conjugation properties of carbon-carbon double bond-linked COF, the photocatalytic yield of Por-sp^2 COF in the process of catalyzing the oxidation of amines to imines can reach 99% within 30 min. Compared with the COF of the imine bond, it shows obviously higher photocatalytic activity and stability. Subsequently, the fully conjugated sp^2-carbon-linked Py-BSZ-COF with donor-acceptor structure was synthesized by the reaction of 1,3,6,8-tetrakis(4-formylphenyl)pyrene and 4,4′-(benzothiadiazole-4,7-diyl)diacetonitrile [50]. The resultant COF has high stability and

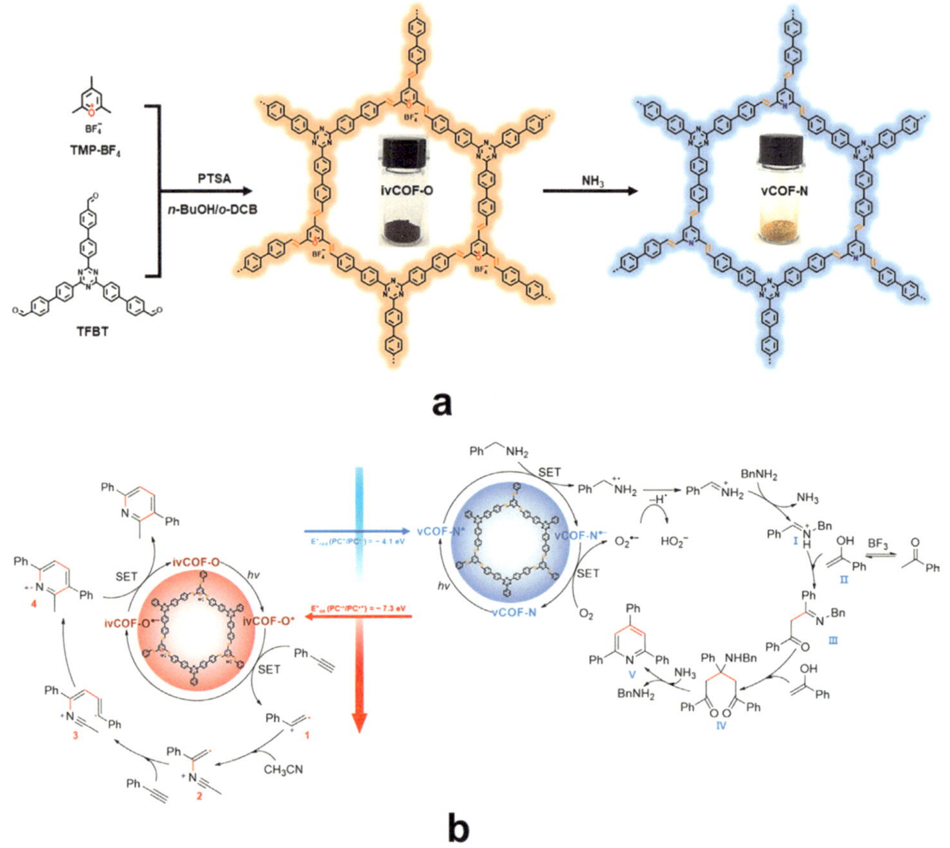

Figure 7.
(a) Synthetic scheme of ivCOF-O and vCOF-N. (b) Photocatalytic organic reactions catalyzed by ivCOF-O and vCOF-N. Adapted with permission from Ref. [44]. Copyright 2021, Wiley-VCH.

can maintain the original skeleton structure after 3 days in 12 mol l^{-1} HCl, 12 mol l^{-1} NaOH, boiling water, and under 15 W white LED light, respectively. The donor-acceptor structure improves the separation ability of photogenerated electrons and holes, making the COF usable for mediating the photocatalytic oxidative amine coupling and cyclization of thioamide to 1,2,4-thiadiazole, with a broad substrate scope and good cyclability.

In 2020, Zhang et al. developed a variety of vinylene-linked COFs with fully sp^2-carbon-connected skeletons by reacting tricyanomesitylene with multi-topic aromatic aldehydes through Knoevenagel condensation (**Figure 7**) [51]. With the use of appropriate secondary amines as catalysts, which could generate highly reactive iminium cation intermediates by reacting with aldehyde groups, the reversibility of the Knoevenagel condensation was greatly enhanced and highly crystalline vinylene-bridged COFs were achieved, which exhibited long-range ordered skeletons, well-defined nanopores, high surface areas (up to 1231 $m^2\ g^{-1}$), efficient light-harvesting capabilities, and facilitated exciton migration and charge transport. Upon visible-light irradiation, these COFs accelerated oxidative hydroxylation of arylboronic acids to phenols with a low loading amount and short reaction time, comparable to homogeneous catalysts (e.g., $Ru(bpy)_3Cl_2$, methylene blue), but recyclable.

Very recently, Müllen et al. developed a 2D vinylene-linked COF comprising nanographene units by employing a dibenzo[*hi*,*st*]ovalene-based building block to react with 3,5-dicyano-2,4,6-trimethylpyridine (DCTMP) [52]. The resultant DBOV-COF formed unique ABC-stacked sp^2-carbon lattices with robust vinylene linkages. By virtue of the narrow-energy-gap nanographene cores as active sites, DBOV-COF demonstrated high photoconductivity, enhanced charge-carrier mobility, and remarkable photocatalytic activity in hydroxylation of boronic acid to phenols with stable cycling performance. This work illustrated a facile strategy of introducing various functional building blocks (e.g., nanographenes) to the library of COFs to novel structures and promising properties (e.g., enhanced photocatalytic activity).

In addition to the photocatalytic oxidative reaction, vinylene-linked COFs have also been explored as efficient photocatalysts for much more complicated photoredox reaction systems. In 2021, Zhang et al. reported a two-dimensional pyrylium-based COF with vinylene linkages through aldol condensation of a trimethylsubstituted pyrylium salt with a tritopic aromatic aldehyde [53]. The resultant oxonium-embedded 2D vinylene-linked COF, namely ivCOF-O, was further converted to a neutral pyridine-cored COF (vCOF-N) by in situ replacement of oxonium ions with nitrogen atoms under a post-synthetic ammonia treatment. Both COFs are conceptually isoelectronic with each other, which means that they have similar geometric structures but significantly different electronic structures. The band structures of the two COFs were evaluated by UV-vis DRS and UPS analyses, showing the ivCOF-O with high oxidative potential and vCOF-N with low reductive potential. Consequently, ivCOF-O enabled photocatalytic [2 + 2 + 2] cyclization of alkynes with nitriles to pyridines upon visible light irradiation due to its high oxidative potential that induced the single electron transfer from alkyne to pyrilium. On the other hand, the lower reductive potential of vCOF-N allows it to reduce molecular oxygen to superoxide radical anion under the visible light excitation. Then the generated active oxygen species induced the coupling of amine with ketone to produce 2,4,6-triarylpyridine derivatives with good substrate tolerance and recyclability. This work provides a method for the precise construction of heteroatom-embedded COFs with highly tailorable band structures and energy levels, thus prospectively expanding the practical applications of COF materials, especially in photocatalysis.

Chen et al. constructed BBO-COF by self-condensation of monomer containing aldehyde and amino functional groups [54]. This "two-in-one" strategy can facilitate the construction of imine-bonded COF with the benzoxazole linkage. The COF can maintain stability in 12 mol l^{-1} HCl and 12 mol l^{-1} NaOH. BBO-COF showed high activity, good substrate tolerance, and reusability in the process of photocatalytic oxidative hydroxylation of aryl boric acid. After 10 cycles, COF still maintains the original structure and catalytic activity.

Using the post-synthetic modification (PSM) strategy to transform reversible bonds into irreversible bonds in COFs is also an effective way to enhance the stability of COFs. Wang et al. constructed a series of benzoxazole-linked COFs with the strategy of "killing two birds with one stone" [55]. After 3 days of storage in boiling water, 9 mol l^{-1} HCl, and 9 mol l^{-1} NaOH, the COFs' crystallinity is still retained, indicating the ultrahigh stability of these COFs. The benzoxazole units introduced by means of PSM into these conjugated COF skeletons can not only effectively improve the light absorption capacity but also provide significant catalytic sites. The yield of this kind of COF in the photocatalytic oxidative hydroxylation of aryl boric acid can reach 99%. Meanwhile, it has high recyclability, which was confirmed by that the skeletal structure and catalytic activity are still maintained after 20 cycles.

The morphology of COFs also has a significant influence on their photocatalytic activity. Aleman et al. synthesized a series of imine-linked COFs with spherical, layered, and 3D structures, respectively [56]. They used these COFs to study the effects of molecular structure, crystallinity, and microstructure on photocatalytic performance. In the selective oxidation of thioether to sulfoxide, crystallinity has a great influence on the photocatalytic activity of 2D COF with spherical and layered morphology, while it has a small influence on 3D COF.

In addition to the abovementioned reactions in photoredox organic transformations catalyzed by COFs, several other reactions, such as the polymerization of methyl methacrylate, Alder–Ene reaction, and *trans* (E) to *cis* (Z) isomerization of olefins, etc. have also been exploited by employing COFs as the photocatalysts. It showcased that COFs hold great potential in photoredox organic transformation yet need to be further explored.

5. Photocatalytic degradation of pollutants

In recent years, with the development of industrialization, the natural environment has been greatly affected, and a variety of different types of pollutants need to be dealt with. Photocatalytic degradation of pollutants is an effective way to solve environmental pollution. In this process, singlet oxygen $^1O_2$ is usually induced by photocatalysts to kill pathogenic bacteria in the environment. In addition, photo-excited catalysts can be used to generate electrons and holes. The photogenerated electrons reduce O_2 to $\cdot O_2^-$ and oxidize OH^- to $\cdot OH^-$ in the environment. Both active oxygen species can be used to degrade the pollutants in the environment.

5.1 Singlet oxygen generated by COFs for catalysis

Porphyrins and their derivatives can generate excited triplet states under light irradiation and then interact with oxygen molecules to form singlet oxygen. Heterogeneous catalytic singlet oxygen formation can be achieved by introducing porphyrin derivatives into COFs. CuP-SQ-COF was synthesized by Jiang et al. using copper porphyrin derivatives and squaric acid [57]. The squaraine-linked COFs have a zigzagged conformation that protects the layered structure from sideslip, are highly stable in solvents, provide an extended π-conjugation over the 2D sheets, and have lower band gap energy and greatly enhanced absorbance capability. Using 1,3-diphenylisobenzofuran (DPBF) as a label, CUP-SQ-COF can effectively produce singlet oxygen under visible light irradiation.

Hydrogen-bonding interactions in the COF intralayer can effectively enhance the stability and photocatalytic activity. Jiang et al. reported the condensation between 5,10,15,20-tetrakis(4′-tetraphenylamino) porphyrin derivatives with terephthaladehyde and dihydroxyterephthalaldehyde in different proportions to obtain a series of COF with different numbers of intramolecular hydrogen bonds, which were used in photocatalysis to produce singlet oxygen (**Figure 8**) [58]. The intramolecular hydrogen bonds formed between imine bonds and hydroxyl groups lock the conformation of 2D in-plane fragments, thereby enhancing the interlayer interaction of COF and enabling it to generate excited triplet states faster and to trigger the transformation of oxygen molecules. The results show that with the increase of hydrogen bonds in COFs, the photocatalytic singlet oxygen generation rate increases gradually. In addition to porphyrin-based COF, imine-linked COFs containing triazine units can

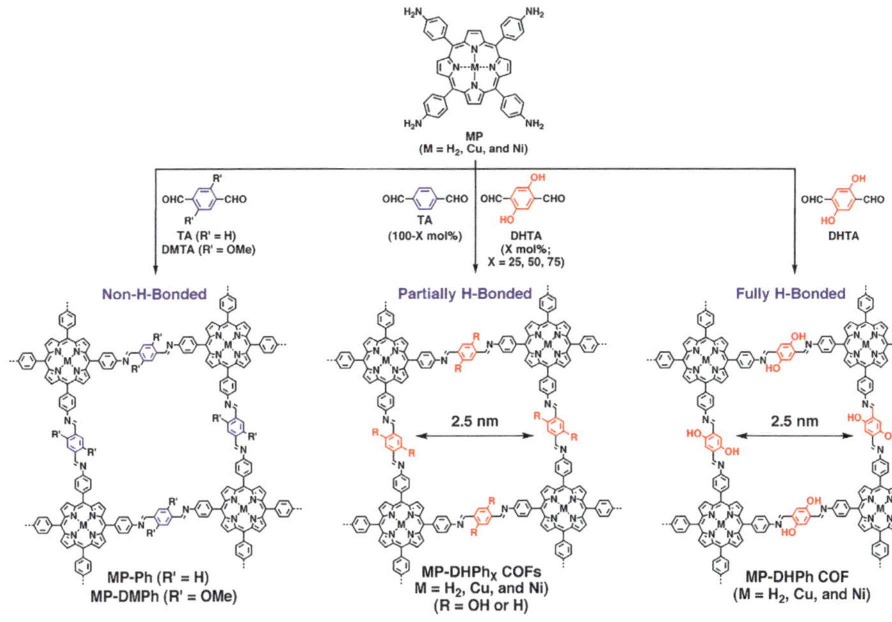

Figure 8.
Schematic of the synthesis of 2D porphyrin COFs with tunable content of hydrogen-bonding structures. Adapted with permission from Ref. [58]. Copyright 2015, American Chemical Society.

also be used for photocatalytic production of singlet oxygen. Zhang et al. synthesized COFs-Trif-Benz and COF-SDU1 by condensation of 2,4,6-tris(4-formylphenyl)-1,3,5-triazine (TFPT) with benzidine and *p*-phenylenediamine, respectively [59]. These two π-conjugated imine COFs can be used to produce singlet oxygen, and then their bactericidal effect has been explored. After 60–90 min of visible light irradiation, 90% of grape mold and Escherichia coli were killed.

5.2 Electrons and holes generated by COFs for catalysis

Organic dyes are usually toxic and carcinogenic pollutants. Photocatalytic degradation of organic dyes into harmless substances such as CO_2 and water is of great value. Cai et al. used 1,3,5-triformylphloroglucinol to condense with melamine to form imine-bonded COF, and then the imine bonds irreversibly tautomerize to obtain TpMA COF, which is stable in acid, base, and under light [60]. Upon visible light irradiation, TpMA COF can generate electrons and holes to reduce O_2 to $\cdot O_2^-$ and oxidize OH^- to $\cdot OH^-$, respectively. Since $\cdot O_2^-$ and $\cdot OH^-$ have strong oxidation ability, the COF can effectively catalyze photodegradation of methyl orange dye with apparent kinetic constant of 0.102 min^{-1}, which is 59 times that of g-C_3N_4.

In order to further enhance the photocatalytic degradation ability of COFs, it is an effective strategy to construct *Z*-scheme heterojunction catalysts based on COFs. *Z*-scheme heterojunctions can effectively inhibit the recombination of photogenerated carriers and thus improve the ability of photocatalytic degradation of pollutants. Cai et al. combined NH_2-MIL-(Ti) with 4,4′,4″-(1,3,5-triazine-2,4,6-triyl)tribenzaldehyde (TTB) and 4,4′,4″-(1,3,5-triazine-2,4,6-triyl)trianiline (TTA) to synthesize MOF/COF composites by one-pot method [61]. The *Z*-scheme heterojunction formed by NH_2-MIL-(Ti) and TTB-TTA COF can enhance the separation and migration

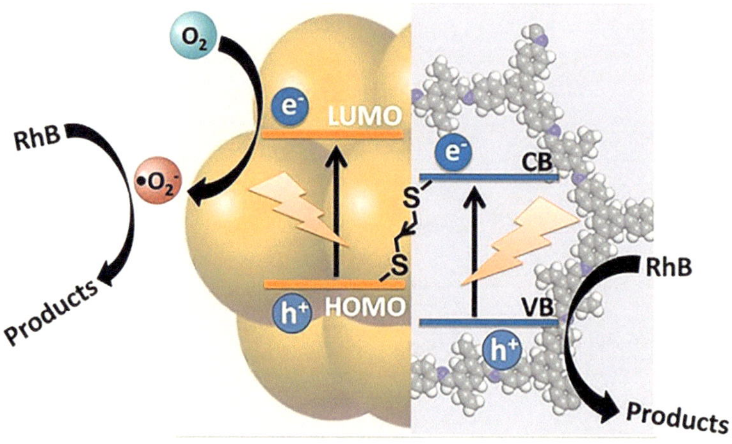

Figure 9.
Principles of photogenerated electron transport between Au NCs and the COF support. Adapted with permission from Ref. [62]. Copyright 2020, Wiley-VCH.

ability of photogenerated carriers and improve the photocatalytic degradation performance of organic pollutants. Under visible light irradiation, photogenerated electrons migrate from the conduction band of NH_2-MIL-(Ti) to the valence band of TTB-TTA COF, generating $\cdot OH^-$ at the valence band of NH_2-MIL-(Ti) and $\cdot O_2^-$ at the conduction band of TTB-TTA COF, which are used to degrade methyl orange dye. The photodegradation kinetics of methyl orange by NH_2-MIL-(Ti)/TTB-TTA COF heterojunction were nine times and two times that of pure NH_2-MIL-(Ti) and pure TTB-TTA COF, respectively. Noble metal nanoparticles can also form heterojunctions with COF for photocatalytic pollutant degradation. Lu et al. anchored gold nanoparticles with thiol-modified imine-bonded COF to form *Z*-scheme heterojunction, thus improving the efficiency of charge separation (**Figure 9**) [62]. Under visible light irradiation, the degradation rate of rhodamine dye and bisphenol A by Au-S-COF within 30 min could reach 93.7% and 90%, respectively. In addition, the degradation ability of Au-S-COF was tested in a continuous flow system contaminated by rhodamine dye. In the flow device, Au-S-COF was used as the filter membrane, the clear solution could directly flow out from the end of the filter. The used Au-S-COF filter paper changed from red to yellow after 30 min of light irradiation, indicating that the pollutant could be completely degraded by light.

In addition to the degradation of organic pollutants, the photogenerated electrons of COF can also be used to reduce heavy metal ions in the environment. Chen et al. used tris(4-aminophenyl)benzene (TPB) or tris-(4-aminophenyl)triazine (TAPT) as donor and electron-deficient benzothiadiazole (BT) as acceptor to construct two COFs with donor-acceptor structure, which were denoted as TPB-BT-COF and TAPT-BT-COF, respectively [63]. The Mott-Schottky spectra of TPB-BT-COF and TAPT-BT-COF show that their conduction band positions are −0.44 and −0.30 V (vs. NHE), respectively. Both are lower than the reduction potential of hexavalent chromium to less-toxic trivalent chromium (1.33 V vs. NHE). This indicated that the COFs can be used to reduce chromium heavy metal ions by photocatalysis. The reduction rate of hexavalent chromium can reach 99% without adding any sacrificial agent or adjusting the pH value. Such high performance was attributable to the stronger photoactivity of TPB-BT-COF due to its more negative band position and narrower band gap.

COFs as purely organic materials might have the smallest impact on the natural environment and have great promise in the practical application of environmental remediation. Thus, one could expect that massive efforts would be made to advance the COF photocatalysts in photocatalytic pollutant degradation.

6. Summary and outlook

As a new type of crystalline organic porous materials, COFs have attracted the attention of researchers due to their unique structural and property advantages and have been rapidly developed in the application of photocatalysis. The designability of the building block makes COFs predictable in structure and adjustable in function, so as to achieve strong light absorption ability and superior stability. The ordered structure and uniform channel can improve the separation and migration ability of photogenerated carriers. High specific surface area can expose more catalytic sites and enhance catalytic activity. All these make COFs become advanced heterogeneous photocatalysts with relatively high conversion rates and excellent recyclability.

In recent years, the research on COF photocatalysts has made some achievements, but it is still in the stage of rapid development. At the same time, there are still some challenges in the field of photocatalytic application by using COFs:(1) rapid and scalable synthesis of COFs photocatalysts. At present, the methods of synthesizing high-quality COFs are cumbersome and time-consuming, which restrict the development of their photocatalytic research and industrial application. (2) Since the photocatalytic mechanism of COFs needs to be further explored, it is a great challenge to design the specific COFs photocatalysts according to the requirements of the specific catalytic reactions. (3) Currently, the number of photocatalytic cycles catalyzed by COFs is limited, and the development of highly stable COFs is an important issue to expand the application scope of COFs photocatalysts. (4) The structure of photoactive COFs is simple. Light absorptive monomers can effectively increase light absorption through covalent bonding to the π-conjugated COFs skeleton. However, suitable photoactive monomers are limited. The design of reasonable photoactive monomers is an effective method to enhance the photosensitivity of COFs. (5) When COFs are used in photocatalysis, most of the reaction systems need additional noble metal cocatalysts and sacrificial agents. It is an important development direction to design COFs photocatalytic systems that are independent of cocatalysts and sacrificial agents. (6) The preparation of high-efficiency and low-cost COFs photocatalytic devices will provide an important way to address the environmental pollution issue and energy crisis.

Acknowledgements

The author is profoundly thankful to the National Natural Science Foundation of China for financial support within the project (22005189) and the China Postdoctoral Science Foundation (2020M681277).

Conflict of interest

The author declares no conflict of interest.

Author details

Shuai Bi
School of Chemistry and Chemical Engineering, Shanghai Jiao Tong University, Shanghai, China

*Address all correspondence to: bishuai-kiven@sjtu.edu.cn

References

[1] Diercks CS, Yaghi OM. The atom, the molecule, and the covalent organic framework. Science. 2017;**355**:eaal1585. DOI: 10.1126/science.aal1585

[2] Diercks CS, Kalmutzki MJ, Yaghi OM. Covalent organic frameworks-organic chemistry beyond the molecule. Molecules. 2017;**22**:1575. DOI: 10.3390/molecules22091575

[3] Feng X, Ding X, Jiang D. Covalent organic frameworks. Chemical Society Reviews. 2012;**41**:6010-6022. DOI: 10.1039/C2CS35157A

[4] Ding S-Y, Wang W. Covalent organic frameworks (COFs): From design to applications. Chemical Society Reviews. 2013;**42**:548-568. DOI: 10.1039/C2CS35072F

[5] Côté AP, Benin AI, Ockwig NW, O'Keeffe M, Matzger AJ, Yaghi OM. Porous, crystalline, covalent organic frameworks. Science. 2005;**310**:1166-1170. DOI: 10.1126/science.1120411

[6] Li Y, Chen W, Xing G, Jiang D, Chen L. New synthetic strategies toward covalent organic frameworks. Chemical Society Reviews. 2020;**49**:2852-2868. DOI: 10.1039/D0CS00199F

[7] Geng K, He T, Liu R, Dalapati S, Tan KT, Li Z, et al. Covalent organic frameworks: Design, synthesis, and functions. Chemical Reviews. 2020;**120**:8814-8933. DOI: 10.1021/acs.chemrcv.9b00550

[8] Liu R, Tan KT, Gong Y, Chen Y, Li Z, Xie S, et al. Covalent organic frameworks: An ideal platform for designing ordered materials and advanced applications. Chemical Society Reviews. 2021;**50**:120-242. DOI: 10.1039/D0CS00620C

[9] Bi S, Meng F, Zhang Z, Wu D, Zhang F. Covalent organic frameworks with *trans*-dimensionally vinylene-linked π-conjugated motifs. Chemical Research in Chinese Universities. 2022;**38**:382-395. DOI: 10.1007/s40242-022-2010-4

[10] Guan X, Chen F, Fang Q, Qiu S. Design and applications of three dimensional covalent organic frameworks. Chemical Society Reviews. 2020;**49**:1357-1384. DOI: 10.1039/C9CS00911F

[11] Huang N, Wang P, Jiang D. Covalent organic frameworks: A materials platform for structural and functional designs. Nature Reviews Materials. 2016;**1**:16068. DOI: 10.1038/natrevmats.2016.68

[12] Wang H, Wang H, Wang Z, Tang L, Zeng G, Xu P, et al. Covalent organic framework photocatalysts: Structures and applications. Chemical Society Reviews. 2020;**49**:4135-4165. DOI: 10.1039/D0CS00278J

[13] Sharma RK, Yadav P, Yadav M, Gupta R, Rana P, Srivastava A, et al. Recent development of covalent organic frameworks (COFs): Synthesis and catalytic (organic-electro-photo) applications. Materials Horizons. 2020;**7**:411-454. DOI: 10.1039/C9MH00856J

[14] Banerjee T, Gottschling K, Savasci G, Ochsenfeld C, Lotsch BV. H_2 evolution with covalent organic framework photocatalysts. ACS Energy Letters. 2018;**3**:400-409. DOI: 10.1021/acsenergylett.7b01123

[15] Linsebigler AL, Lu G, Yates JT. Photocatalysis on TiO_2 surfaces: Principles, mechanisms, and selected

results. Chemical Reviews. 1995;**95**:735-758. DOI: 10.1021/cr00035a013

[16] Vyas VS, W-h LV, Lotsch BV. Soft photocatalysis: Organic polymers for solar fuel production. Chemistry of Materials. 2016;**28**:5191-5204. DOI: 10.1021/acs.chemmater.6b01894

[17] Stegbauer L, Schwinghammer K, Lotsch BV. A hydrazone-based covalent organic framework for photocatalytic hydrogen production. Chemical Science. 2014;**5**:2789-2793. DOI: 10.1039/C4SC00016A

[18] Vyas VS, Haase F, Stegbauer L, Savasci G, Podjaski F, Ochsenfeld C, et al. A tunable azine covalent organic framework platform for visible light-induced hydrogen generation. Nature Communications. 2015;**6**:8508. DOI: 10.1038/ncomms9508

[19] Haase F, Banerjee T, Savasci G, Ochsenfeld C, Lotsch BV. Structure-property-activity relationships in a pyridine containing azine-linked covalent organic framework for photocatalytic hydrogen evolution. Faraday Discussions. 2017;**201**:247-264. DOI: 10.1039/C7FD00051K

[20] Stegbauer L, Zech S, Savasci G, Banerjee T, Podjaski F, Schwinghammer K, et al. Tailor-made photoconductive pyrene-based covalent organic frameworks for visible-light driven hydrogen generation. Advanced Energy Materials. 2018;**8**:1703278. DOI: 10.1002/aenm.201703278

[21] Pachfule P, Acharjya A, Roeser J, Langenhahn T, Schwarze M, Schomäcker R, et al. Diacetylene functionalized covalent organic framework (COF) for photocatalytic hydrogen generation. Journal of the American Chemical Society. 2018;**140**:1423-1427. DOI: 10.1021/jacs.7b11255

[22] Wang X, Chen L, Chong SY, Little MA, Wu Y, Zhu W-H, et al. Sulfone-containing covalent organic frameworks for photocatalytic hydrogen evolution from water. Nature Chemistry. 2018;**10**:1180-1189. DOI: 10.1038/s41557-018-0141-5

[23] Sprick RS, Bonillo B, Clowes R, Guiglion P, Brownbill NJ, Slater BJ, et al. Visible-light-driven hydrogen evolution using planarized conjugated polymer photocatalysts. Angewandte Chemie International Edition. 2016;**55**:1792. DOI: 10.1002/anie.201510542

[24] Sheng J-L, Dong H, Meng X-B, Tang H-L, Yao Y-H, Liu D-Q, et al. Effect of different functional groups on photocatalytic hydrogen evolution in covalent-organic frameworks. ChemCatChem. 2019;**11**:2313-2319. DOI: 10.1002/cctc.201900058

[25] Bi S, Yang C, Zhang W, Xu J, Liu L, Wu D, et al. Two-dimensional semiconducting covalent organic frameworks via condensation at arylmethyl carbon atoms. Nature Communications. 2019;**10**:2467. DOI: 10.1038/s41467-019-10504-6

[26] Wei S, Zhang W, Qiang P, Yu K, Fu X, Wu D, et al. Semiconducting 2D triazine-cored covalent organic frameworks with unsubstituted olefin linkages. Journal of the American Chemical Society. 2019;**141**:14272-14279. DOI: 10.1021/jacs.9b06219

[27] Xu J, Yang C, Bi S, Wang W, He Y, Wu D, et al. Vinylene-linked covalent organic frameworks (COFs) with symmetry-tuned polarity and photocatalytic activity. Angewandte Chemie International Edition. 2020;**59**:23845-23853. DOI: 10.1002/anie.202011852

[28] Jin E, Lan Z, Jiang Q, Geng K, Li G, Wang X, et al. 2D sp^2 carbon-conjugated covalent organic frameworks for photocatalytic hydrogen production from water. Chem. 2019;**5**:1632-1647. DOI: j.chempr.2019.04.015

[29] Luo M, Yang Q, Liu K, Cao H, Yan H. Boosting photocatalytic H_2 evolution on g-C_3N_4 by modifying covalent organic frameworks (COFs). Chemical Communications. 2019;**55**:5829-5832. DOI: 10.1039/C9CC02144B

[30] Zhang F-M, Sheng J-L, Yang Z-D, Sun X-J, Tang H-L, Lu M, et al. Rational design of MOF/COF hybrid materials for photocatalytic H_2 evolution in the presence of sacrificial electron donors. Angewandte Chemie International Edition. 2018;**57**:12106-12110. DOI: 10.1002/anie.201806862

[31] Thote J, Aiyappa HB, Deshpande A, Díaz Díaz D, Kurungot S, Banerjee R. A covalent organic framework-cadmium sulfide hybrid as a prototype photocatalyst for visible-light-driven hydrogen production. Chemistry—A European Journal. 2014;**20**:15961-15965. DOI: 10.1002/chem.201403800

[32] Ming J, Liu A, Zhao J, Zhang P, Huang H, Lin H, et al. Hot π-electron tunneling of metal-insulator-COF nanostructures for efficient hydrogen production. Angewandte Chemie International Edition. 2019;**58**:18290-18294. DOI: 10.1002/anie.201912344

[33] Banerjee T, Haase F, Savasci G, Gottschling K, Ochsenfeld C, Lotsch BV. Single-site photocatalytic H_2 evolution from covalent organic frameworks with molecular cobaloxime Co-catalysts. Journal of the American Chemical Society. 2017;**139**:16228-16234. DOI: 10.1021/jacs.7b07489

[34] Biswal BP, Vignolo-González HA, Banerjee T, Grunenberg L, Savasci G, Gottschling K, et al. Sustained solar H_2 evolution from a thiazolo[5,4-d]thiazole-bridged covalent organic framework and nickel-thiolate cluster in water. Journal of the American Chemical Society. 2019;**141**:11082-11092. DOI: 10.1021/jacs.9b03243

[35] Xu R, Wang X-S, Zhao H, Lin H, Huang Y-B, Cao R. Rhenium-modified porous covalent triazine framework for highly efficient photocatalytic carbon dioxide reduction in a solid-gas system. Catalysis Science & Technology. 2018;**8**:2224-2230. DOI: 10.1039/C8CY00176F

[36] Yang S, Hu W, Zhang X, He P, Pattengale B, Liu C, et al. 2D covalent organic frameworks as intrinsic photocatalysts for visible light-driven CO_2 reduction. Journal of the American Chemical Society. 2018;**140**:14614-14618. DOI: 10.1021/jacs.8b09705

[37] Zhong W, Sa R, Li L, He Y, Li L, Bi J, et al. A covalent organic framework bearing single Ni sites as a synergistic photocatalyst for selective photoreduction of CO_2 to CO. Journal of the American Chemical Society. 2019;**141**:7615-7621. DOI: 10.1021/jacs.9b02997

[38] Lu M, Li Q, Liu J, Zhang F-M, Zhang L, Wang J-L, et al. Installing earth-abundant metal active centers to covalent organic frameworks for efficient heterogeneous photocatalytic CO_2 reduction. Applied Catalysis B: Environmental. 2019;**254**:624-633. DOI: 10.1016/j.apcatb.2019.05.033

[39] Lu M, Liu J, Li Q, Zhang M, Liu M, Wang J-L, et al. Rational design of crystalline covalent organic frameworks for efficient CO_2 photoreduction with H_2O. Angewandte Chemie International Edition. 2019;**58**:12392-12397. DOI: 10.1002/anie.201906890

[40] Lin S, Diercks CS, Zhang Y-B, Kornienko N, Nichols EM, Zhao Y, et al. Covalent organic frameworks comprising cobalt porphyrins for catalytic CO_2 reduction in water. Science. 2015;**349**:1208-1213. DOI: 10.1126/science.aac8343

[41] Liu W, Li X, Wang C, Pan H, Liu W, Wang K, et al. A scalable general synthetic approach toward ultrathin imine-linked two-dimensional covalent organic framework nanosheets for photocatalytic CO_2 reduction. Journal of the American Chemical Society. 2019;**141**:17431-17440. DOI: 10.1021/jacs.9b09502

[42] Fu Y, Zhu X, Huang L, Zhang X, Zhang F, Zhu W. Azine-based covalent organic frameworks as metal-free visible light photocatalysts for CO_2 reduction with H_2O. Applied Catalysis B: Environmental. 2018;**239**:46-51. DOI: 10.1016/j.apcatb.2018.08.004

[43] Lei K, Wang D, Ye L, Kou M, Deng Y, Ma Z, et al. A metal-free donor-acceptor covalent organic framework photocatalyst for visible-light-driven reduction of CO_2 with H_2O. ChemSusChem. 2020;**13**:1725-1729. DOI: 10.1002/cssc.201903545

[44] Zhang M, Lu M, Lang Z-L, Liu J, Liu M, Chang J-N, et al. Semiconductor/covalent-organic-framework Z-scheme heterojunctions for artificial photosynthesis. Angewandte Chemie International Edition. 2020;**59**:6500-6506. DOI: 10.1002/anie.202000929

[45] Zhi Y, Li Z, Feng X, Xia H, Zhang Y, Shi Z, et al. Covalent organic frameworks as metal-free heterogeneous photocatalysts for organic transformations. Journal of Materials Chemistry A. 2017;**5**:22933-22938. DOI: 10.1039/C7TA07691F

[46] Li Z, Zhi Y, Shao P, Xia H, Li G, Feng X, et al. Covalent organic framework as an efficient, metal-free, heterogeneous photocatalyst for organic transformations under visible light. Applied Catalysis B: Environmental. 2019;**245**:334-342. DOI: 10.1016/j.apcatb.2018.12.065

[47] Liu W, Su Q, Ju P, Guo B, Zhou H, Li G, et al. A hydrazone-based covalent organic framework as an efficient and reusable photocatalyst for the cross-dehydrogenative coupling reaction of N-aryltetrahydroisoquinolines. ChemSusChem. 2017;**10**:664-669. DOI: 10.1002/cssc.201601702

[48] Bhadra M, Kandambeth S, Sahoo MK, Addicoat M, Balaraman E, Banerjee R. Triazine functionalized porous covalent organic framework for photo-organocatalytic *E-Z* isomerization of olefins. Journal of the American Chemical Society. 2019;**141**:6152-6156. DOI: 10.1021/jacs.9b01891

[49] Chen R, Shi J-L, Ma Y, Lin G, Lang X, Wang C. Designed synthesis of a 2D porphyrin-based sp^2 carbon-conjugated covalent organic framework for heterogeneous photocatalysis. Angewandte Chemie International Edition. 2019;**58**:6430-6434. DOI: 10.1002/anie.201902543

[50] Li S, Li L, Li Y, Dai L, Liu C, Liu Y, et al. Fully conjugated donor-acceptor covalent organic frameworks for photocatalytic oxidative amine coupling and thioamide cyclization. ACS Catalysis. 2020;**10**:8717-8726. DOI: 10.1021/acscatal.0c01242

[51] Bi S, Thiruvengadam P, Wei S, Zhang W, Zhang F, Gao L, et al. Vinylene-bridged two-dimensional covalent organic frameworks via knoevenagel condensation of tricyanomesitylene. Journal of

the American Chemical Society. 2020;**142**:11893-11900. DOI: 10.1021/jacs.0c04594

[52] Jin E, Fu S, Hanayama H, Addicoat MA, Wei W, Chen Q, et al. A nanographene-based two-dimensional covalent organic framework as a stable and efficient photocatalyst. Angewandte Chemie International Edition. 2021;**61**:e202114059. DOI: 10.1002/anie.202114059

[53] Bi S, Zhang Z, Meng F, Wu D, Chen J-S, Zhang F. Heteroatom-embedded approach to vinylene-linked covalent organic frameworks with isoelectronic structures for photoredox catalysis. Angewandte Chemie International Edition. 2021;**61**:e202111627. DOI: 10.1002/anie.202111627

[54] Yan X, Liu H, Li Y, Chen W, Zhang T, Zhao Z, et al. Ultrastable covalent organic frameworks via self-polycondensation of an A2B2 monomer for heterogeneous photocatalysis. Macromolecules. 2019;**52**:7977-7983. DOI: 10.1021/acs.macromol.9b01600

[55] Wei P-F, Qi M-Z, Wang Z-P, Ding S-Y, Yu W, Liu Q, et al. Benzoxazole-linked ultrastable covalent organic frameworks for photocatalysis. Journal of the American Chemical Society. 2018;**140**:4623-4631. DOI: 10.1021/jacs.8b00571

[56] Jiménez-Almarza A, López-Magano A, Marzo L, Cabrera S, Mas-Ballesté R, Alemán J. Imine-based covalent organic frameworks as photocatalysts for metal free oxidation processes under visible light conditions. ChemCatChem. 2019;**11**:4916-4922. DOI: 10.1002/cctc.201901061

[57] Nagai A, Chen X, Feng X, Ding X, Guo Z, Jiang D. A Squaraine-linked mesoporous covalent organic framework. Angewandte Chemie International Edition. 2013;**52**:3770-3774. DOI: 10.1002/anie.201300256

[58] Chen X, Addicoat M, Jin E, Zhai L, Xu H, Huang N, et al. Locking covalent organic frameworks with hydrogen bonds: General and remarkable effects on crystalline structure, physical properties, and photochemical activity. Journal of the American Chemical Society. 2015;**137**:3241-3247. DOI: 10.1021/ja509602c

[59] Liu T, Hu X, Wang Y, Meng L, Zhou Y, Zhang J, et al. Triazine-based covalent organic frameworks for photodynamic inactivation of bacteria as type-II photosensitizers. Journal of Photochemistry and Photobiology B: Biology. 2017;**175**:156-162. DOI: 10.1016/j.jphotobiol.2017.07.013

[60] He S, Rong Q, Niu H, Cai Y. Construction of a superior visible-light-driven photocatalyst based on a C_3N_4 active Centre-photoelectron shift platform-electron withdrawing unit triadic structure covalent organic framework. Chemical Communications. 2017;**53**:9636-9639. DOI: 10.1039/C7CC04515H

[61] He S, Rong Q, Niu H, Cai Y. Platform for molecular-material dual regulation: A direct *Z*-scheme MOF/COF heterojunction with enhanced visible-light photocatalytic activity. Applied Catalysis B: Environmental. 2019;**247**:49-56. DOI: 10.1016/j.apcatb.2019.01.078

[62] Deng Y, Zhang Z, Du P, Ning X, Wang Y, Zhang D, et al. Embedding ultrasmall au clusters into the pores of a covalent organic framework for enhanced photostability and photocatalytic performance. Angewandte Chemie International Edition. 2020;**59**:6082-6089. DOI: 10.1002/anie.201916154

[63] Chen W, Yang Z, Xie Z, Li Y, Yu X, Lu F, et al. Benzothiadiazole functionalized D-A type covalent organic frameworks for effective photocatalytic reduction of aqueous chromium(VI). Journal of Materials Chemistry A. 2019;7:998-1004. DOI: 10.1039/C8TA10046B

Chapter 4

Photocatalysis of Covalent Organic Frameworks

Hui Liu and Yingjie Zhao

Abstract

The development of clean and sustainable energy is gaining attention in light of the current energy crisis and global warming. An ideal way to utilize renewable solar energy is to convert clean energy through photocatalysis. This includes splitting water, reducing CO_2, regenerating coenzymes, etc. Photocatalysis relies heavily on photocatalysts. It has recently become popular to use organic porous polymers in this process. Covalent organic frameworks (COFs), as one of the organic porous polymers, have the characteristics of high crystallinity, porosity, and structural designability that make them perfect platforms for photocatalysis. An overview of recent advances in COF photocatalysts is presented in this chapter. The photocatalytic applications of COFs with different ligation and different structures were first discussed, including photocatalytic hydrogen evolution, CO_2 conversion, coenzyme regeneration, and conventional organic reactions. Finally, conclusions and prospects were provided in the last section.

Keywords: covalent organic frameworks, photocatalysis, photocatalytic hydrogen evolution, photocatalytic CO_2 reduction

1. Introduction

As fossil fuels (e.g. oil, coal, and natural gas) were excessively utilized, energy crisis and climate issues gradually became global concerns [1–3]. Developing "green energy" as a replacement for traditional fossil fuels will be the most efficient solution to these problems. Photocatalysis is a green technology with important application prospects in energy and environment. Photocatalytic reaction refers to the process of semiconductor photocatalysts promoting the conversion of compounds under light conditions, which can effectively convert light energy into chemical energy. On the one hand, photocatalytic degradation of organic pollutants and photocatalytic reduction of CO_2 is helpful to solve environmental problems. On the other hand, photolysis of hydrogen and oxygen in aquatic products can develop new energy for human beings. It should be noted that photocatalyst materials play an important role in the whole photocatalytic process. During the photocatalytic process, electron–hole pairs in the photocatalysts are generated by absorption of photons with higher energy than the band gap. In redox reactions, electrons and holes migrate to the surface, catalyzing the reaction. The first photoelectrochemical water splitting on TiO_2 was reported by Fujishima and Honda in 1972 [4]. Since then, a series of inorganic photocatalysts including TiO_2 [5, 6], cadmium sulphide (CdS) [7, 8], zinc oxide (ZnO) [9] and silver phosphate (Ag_3PO_4) [10] emerged and lead the research over the past several decades. Its practical application was limited, however, by the requirements for UV light, heavy metal toxicity, and photo corrosion. In recent years, using organic polymers like linear conjugated polymers (CPs) [11–14], graphitic carbon nitride (g-C_3N_4) [15–17], conjugated microporous polymers (CMPs) [18–23], covalent organic frameworks (COFs) [24–27] and covalent triazine-based frameworks (CTFs) [28–31] as photocatalysts became hot topics. Among them, COFs have drawn wide attention due to advantages of easy structure design, large surface area, tunable electronic properties and band gaps, and diverse synthesis methods.

COFs are a new type of crystalline organic porous polymer based on covalent bond connection [32]. COFs with two-dimensional (2D) or three-dimensional (3D) networks could be prepared according to the construction units of different building blocks. The unique structures of COFs also bring a couple of important advantages: (1) the structural designability endows them with enhanced visible-light absorption and tunable band structure; (2) the porous structure is conducive to the adsorption of substrates and the transport of products; (3) the strong covalent bonds endow COFs with good stability, which thereby prolongs the lifetime of the photoactive structure; (4) the ordered conjugated structure is beneficial not only to the absorption of light energy but also to the transport of excited electrons. As a result of these potential advantages, COFs have been extensively used in gas storage and separation [33–36], catalysis [37–39], optoelectronics [40], sensing [41–43], and energy storage [44, 45]. There have been many excellent reviews about the synthesis, structures, and applications of COFs materials [46–55]. It can be said that COFs material has many advantages as a heterogeneous photocatalyst.

In this chapter, we will focus on the photocatalytic applications of COFs including water splitting, CO_2 reduction, coenzyme regeneration, and photocatalytic organic reactions. Finally, we will discuss the challenges and opportunities of COFs as photocatalysts.

2. Photocatalytic applications of COFs

2.1 Photocatalytic hydrogen evolution

Hydrogen energy is regarded as the most promising clean energy in the 21st century because the only product of hydrogen combustion is water [56]. Hydrogen production from powder photocatalyst is expected to break the cost barrier and become the cheapest technology to decompose water to produce hydrogen, which is expected to surpass fossil fuels. In recent years, organic semiconductor materials have been extensively explored in photocatalytic hydrogen evolution. Among them, COFs with well-ordered conjugate structures showed great advantages not only in photocatalytic performance but also in the deep understanding of the structure–activity relationship [55–58]. A great deal of research work has focused on improving the performance of the photocatalyst by adjusting the building units and linkages of the COFs. In this section, representative building blocks and linkages of COFs for photocatalytic hydrogen evolution were summarized and discussed.

2.1.1 Hydrazone-linked COFs

Hydrazone-linked COFs with hydrolytic and oxidative stability were prepared by condensation between hydrazide and aldehyde derivatives. Therefore, hydrazone-linked COFs provide a valuable design platform for water-splitting photocatalysts. In 2014, Lotsch group constructed a 2D COF (TFPT-COF) with hydrazone-linked through the condensation reaction of 2,5-diethoxyterephthalohydrazide and triazine-based aldehyde (**Figure 1**) [25]. This is the first report of COFs materials applied to photocatalytic hydrogen production. TFPT-COF is a mesoporous material with a pore size of 3.8 nm, a specific surface area of 1603 m^2/g, and a pore volume of 1.03 cm^3/g. Compared with monomer, TFPT-COF has better visible light absorption performance, and the absorption boundary can reach more than 600 nm. TFPT-COF was selected as a photosensitizer, while Pt was used as the proton reduction catalyst. When illuminated with visible light, the hydrogen evolution rate in the first five hours reached 1970 μmol h^{-1} g^{-1} using 10 vol% aqueous triethanolamine (TEOA) solution as the sacrificial electron donor. The quantum efficiency was determined to be 2.2%. Under the same conditions, the hydrogen production efficiency of TFPT-COF was much superior to Pt-modified amorphous melon, g-C_3N_4, and crystalline poly (triazine imide) [59].

As a type of conjugated macrocycle, porphyrin displays unique photophysical and redox characteristics [60]. Wang and co-workers designed and synthesized

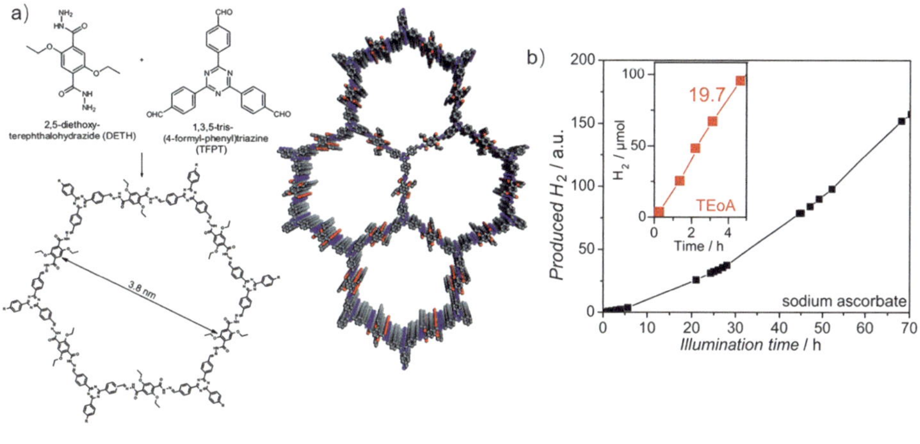

Figure 1.
(a) Synthetic route and top view of the TFPT-COF. (b) Photocatalytic performance of Pt-modified TFPT-COF with different SED (black: sodium ascorbate solution; red: triethanolamine solution).

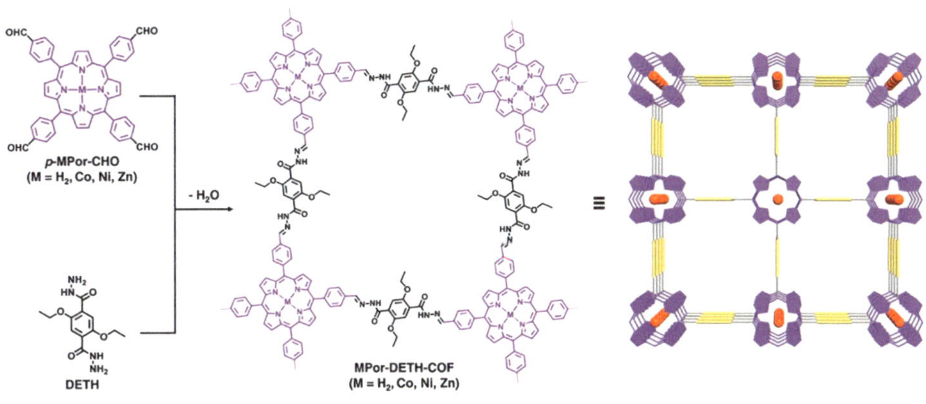

Figure 2.
Schematic representation of the synthesis of MPor-DETH-COFs.

four isostructural porphyrin-based 2D COFs (**Figure 2**) [61]. By incorporating different transition metals into the porphyrin rings, the physical and electronic properties of COFs were rationally tuned. There was a high level of crystallinity and surface area in all of the COFs. When illuminated with visible light while containing Pt as a co-catalyst and TEOA as a sacrificial electron donor, these four COFs exhibited tunable hydrogen evolution efficiency with the order of CoPor-DETH-COF (25 μmol g^{-1} h^{-1}) < H_2Por-DETH-COF (80 μmol g^{-1} h^{-1}) < NiPor-DETH-COF (211 μmol g^{-1} h^{-1}) < ZnPor-DETH-COF (413 μmol g^{-1} h^{-1}). A molecular engineering approach was mainly responsible for the tunable photocatalytic performance. Moreover, hydrazone-linked COFs were also can be converted into oxadiazole-linked COFs via oxidization reaction, which endow COFs with narrow bandgaps and continuous π-electron delocalization for high photocatalytic activity (2615 μmol h^{-1} g^{-1}) [62].

2.1.2 Azine-linked COFs

The azine-linked COFs are synthesized by the condensation of aldehyde derivatives with hydrazine. The electronic band structure and steric hindrance can be regulated by introducing different heteroatoms into COFs. Lostch and co-workers synthesized a series of azine-linked COFs and investigated the connection between the number of nitrogen atoms in their structures and the photocatalytic hydrogen production efficiency (**Figure 3**) [63]. Pt and 10 vol% TEOA were selected as the co-catalyst and sacrificial electron donor, respectively. The photocatalytic performance was gradually enhanced with the increase of nitrogen content in the central benzene ring. The hydrogen production efficiencies of N_0-COF, N_1-COF, N_2-COF, and N_3-COF were 23, 90, 438, and 1703 μmol h^{-1} g^{-1}, respectively. Through theoretical calculation, it is found that the planarity of COFs increased with the increase of nitrogen element content. Therefore, N_3-COF has the best planarity, which was conducive to the photogenerated electron migration and thus improves the corresponding photocatalytic activity. In addition to the above work, the authors also reported a series of planar pyrene-azine-COFs (A-TEXPY-COFs) with varied nitrogen atoms in the peripheral aromatic units [64]. The photocatalytic efficiency of A-TEXPY-COFs was regulated by nitrogen element content in the peripheral aromatic units. The

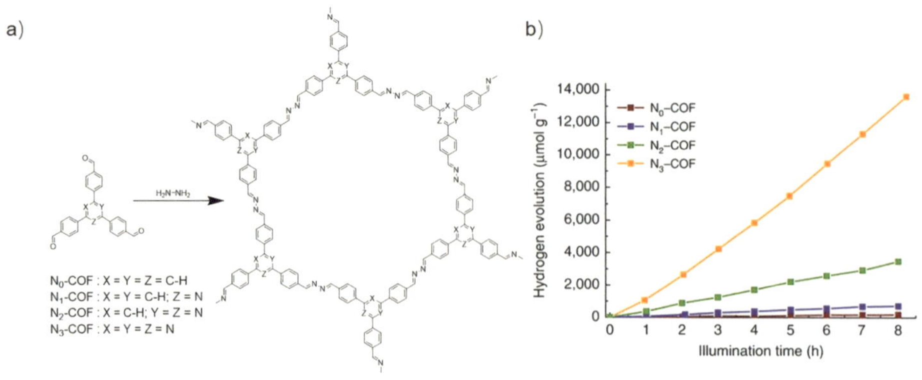

Figure 3.
(a) The formation of N_x-COF. (b) Photocatalytic H_2 evolution of N_x-COFs as photocatalyst over 8 h under visible light irradiation.

photocatalytic efficiency exhibited a decreasing tendency with an increasing nitrogen element content. Quantum-chemical calculations suggested that the thermodynamic driving force increased with the decrease of nitrogen content. In addition, Lostch and co-workers also developed cobaloxime as a co-catalyst instead of platinum, a precious metal [65]. The HER rate of N_2-COF could reach 782 μmol h^{-1} g^{-1} with TEOA as SED.

2.1.3 β-ketoenamine-linked COFs

The β-ketoenamine-linked COFs could be synthesized through Schiff base reaction and this kind of COFs shows great potential in photocatalytic hydrogen production due to its excellent solvent stabilities of keto–enamine. In 2018, Cooper et al. synthesized two COFs (S-COF and FS-COF) with ultra-high photocatalytic hydrogen evolution activity using benzo-bis(benzothiophene sulfone) moieties (**Figure 4**) [27]. Pt acted as the co-catalyst, while the sacrificial electron donor was changed to 0.1 M ascorbic acid. The average photocatalytic HER rate of S-COF and FS-COF were measured to be 4.44 mmol h^{-1} g^{-1} and 10.1 mmol h^{-1} g^{-1}, respectively. Both S-COF and FS-COF have sufficient stability under visible light radiation, and FS-COF could work stably for 50 hours under ascorbic acid conditions. It was noted that HER rate of 1.32 mmol h^{-1} g^{-1} for FS-COF was obtained even in the absence of Pt. Moreover, higher HER of 16.1 and 16.3 mmol h^{-1} g^{-1} were achieved when FS-COF were filled with Eosin Y and WS5F, respectively.

With the introduction of functional organic units in COFs, tunable platforms have become possible. Acetylene as functional group has received significant attention in the field of photocatalysis. For instance, the β-ketoenamine-linked COFs (TP-EDDA and TP-BDDA) were modified with acetylene and diacetylene groups by Thomas et al. (**Figure 5**) [66]. TP-BDDA has a narrower band gap than TP-EDDA due to its enhanced conjugated structure. Therefore, TP-BDDA showed better activity of photocatalytic hydrogen evolution. With Pt as co-catalyst and 10% TEOA as sacrificial electron donor, TP-BDDA exhibited an HER of 324 μmol h^{-1} g^{-1}, substantially higher than that of TP-EDDA (30 μmol h^{-1} g^{-1}). In addition, the optical and electronic properties of β-ketoenamine-linked COFs could be adjusted by slightly changing the structure of their building blocks. Seki and co-workers synthesized a series of β-ketoenamine-linked isoreticular COFs based on 1,3,5-triformylphloroglucinol and 4,4″-diamino-substituted p-terphenyl derivatives [67]. Among these COFs,

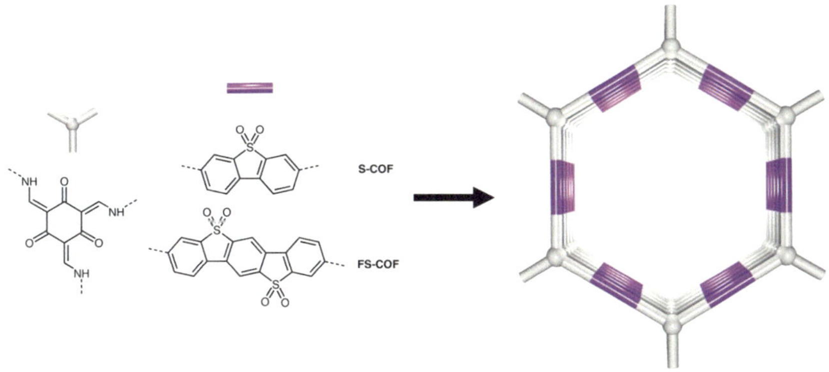

Figure 4.
Synthetic routes for FS-COF and S-COF.

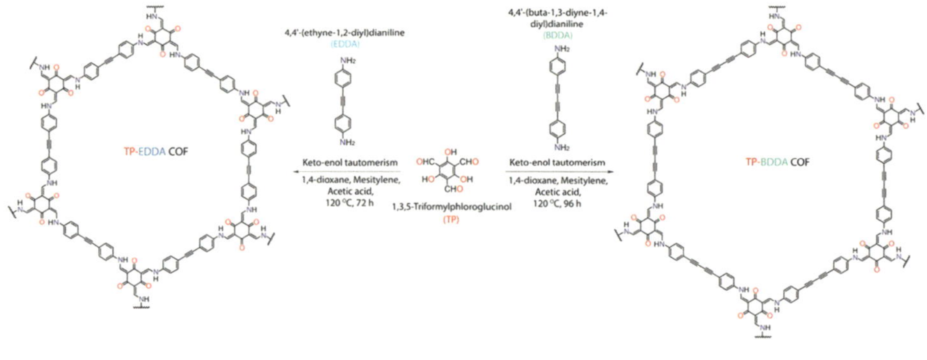

Figure 5.
Scheme of the synthesis of TP-EDDA and TP-BDDA COFs.

electron-deficient benzothiadiazole substituted COF (BtCOF150) exhibited better hydrogen evolution efficiency of 750 μmol h^{-1} g^{-1}.

2.1.4 Imine-linked COFs

In COF synthesis, amine linkages are commonly formed by condensation between aldehydes and amines, which are abundant and easily accessible. Various attempts have been made to construct imine-linked COFs for high photocatalytic activity since the 2010 report on photocatalytic H_2 production from conjugated poly(azomethine) networks [68].

The introduction of D-A strategy into COFs was beneficial to enhance charge separation and transport ability and thus enhaning photocatalytic activity. In 2020, Wen and co-workers constructed PyTz-COF with a band gap of 2.20 eV using the electron-rich pyrene (Py) and electron-deficient thiazolo[5,4-d]thiazole (Tz) [69]. Overlapping orbitals between Py and Tz units enabled electron transfer and charge separation. With ascorbic acid as SED and Pt as co-catalyst, the hydrogen production rate of PyTz-COF was up to 2072.4 μmol h^{-1} g^{-1}. Besides, Dong et al. prepared a benzothiadiazole (BT)-based COF (BT-TAPT-COF) using the same strategy [70]. For at least 64 hours, BT-TAPT-COF demonstrated efficient and steady photocatalytic H_2 evolution. The maximum HER rate was 949 μmol h^{-1} g^{-1} with SED (ascorbic acid) and co-catalyst (Pt). In addition, Chen and co-workers synthesized BT-based COFs that are efficient photocatalysts by introducing halogen moieties into D-A arrangements (**Figure 6**) [71]. It was found that Py-ClTP-BT-COF with co-catalyst (Pt) and SED (ascorbic acid) has superior HER rate of 8875 μmol h^{-1} g^{-1}. At 420 nm, Py-ClTP-BT-COF achieved an apparent quantum efficiency (AQE) of 8.45%, which was higher than most reported COF-based photocatalysts at that time. Besides, the HER rate of Py-ClTP-BT-COF could also reach 2200 μmol h^{-1} g^{-1} without co-catalyst.

Although imine-linked COFs show the potential of photocatalytic hydrogen production, but their stability and conjugation degree are still their shortcomings, which also restricts the further improvement of their photocatalytic efficiency.

2.1.5 Vinylene-linked COFs

It has been recently discovered that COFs with vinylene-linked have been fabricated via knoevenagel condensation or aldol condensation. As compared to imine, hydrazone, and azine-linked COFs, vinylene-linked COFs show greater conjugation

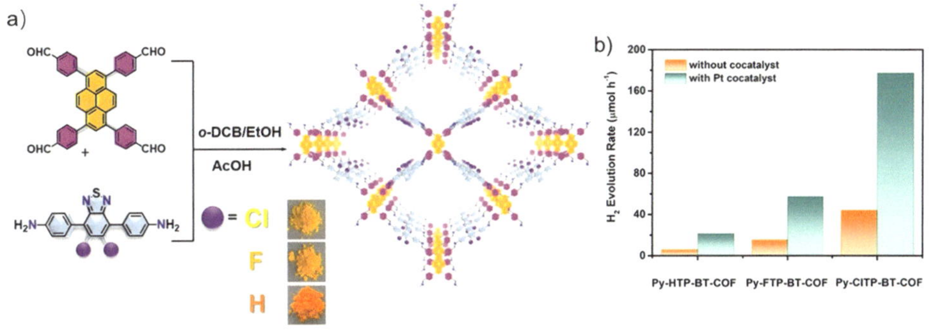

Figure 6.
(a) The formation of Py-XTP-BT-COFs and stacking mode of Py-ClTP-BT-COF. (b) Photocatalytic hydrogen evolution of Py-XTP-BT-COFs under visible light irradiation.

degree. For vinylene-linked COFs, extended conjugation through C=C linkages not only enhances photothermal stability but also enhances absorbance and exciton migration. It was anticipated that vinylene-linked COFs, which combine chemical stability with crystallinity, porosity, and the ability to conjugate, would be highly photoactive. Although the vinylene-linked has many advantages, the poor reversibility of this linkage makes the construction of vinylene-linked very challenging because the bad self-adjusting process is unfavorable for the formation of highly ordered structures.

Jiang and co-workers first synthesized sp^2 carbon frameworks by knoevenagel condensation. Then, a strong electron-withdrawing group (3-ethylrho-danine, ERDN) was introduced at the edge of sp^2c-COF to obtain D-A heterojunction (**Figure 7**) [72]. In consequence, visible light was clearly absorbed more widely. As a result of the push-and-pull effect, exciton migration and charge transport were also facilitated. When illuminated with visible light in the presence of Pt and TEOA, sp^2 c-COF_{ERDN} exhibited an HER of 2.12 mmol g^{-1} h^{-1}, higher than sp^2 c-COF (1.36 mmol g^{-1} h^{-1}). According to Zhang and co-workers, vinylene-linked COFs have been investigated as effective photocatalytic water splitter [73, 74]. As an example, they designed and synthesized two triazine-cored COFs (g-$C_{18}N_3$-COF and g-$C_{33}N_3$-COF) with unsubstituted olefin linkages [74]. The higher HER rate was observed in g-$C_{18}N_3$-COF (14.6 µmol h^{-1}) and

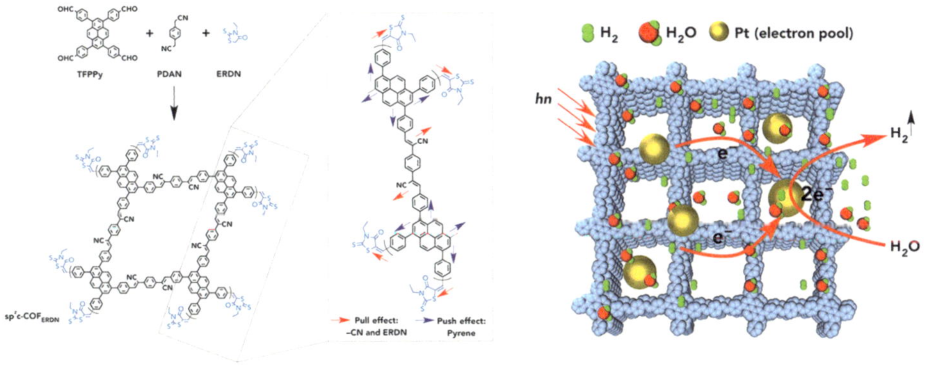

Figure 7.
Schematic representation of the synthesis and photocatalytic hydrogen evolution of sp^2c-COFERDN.

g-$C_{33}N_3$-COF (3.7 μmol) under the same conditions than in imine-linked COF with a similar topology and unit cell parameters. Furthermore, the ordered conjugated structure of g-$C_{18}N_3$-COF yielded a high photocurrent of 45 μA cm^{-2} at 0.2 V vs. RHE.

In order to achieve high photocatalytic performance, effective charge separation and transport must be achieved. Structures built between donors and acceptors are proving to be highly effective. According to Zhao et al., sp^2-carbon-linked COFs with periodic D-A structures were shown in a recent study (**Figure 8**) [75]. Cyano-vinylene linkages are incorporated into an electron-deficient benzobisthiazole structure to form a conjugated structure. It was found that the two-dimensional networks exhibited a strong D-A effect when electron-rich benzotrithiophene was introduced. Taking advantage of the highly conjugated and ordered structure of D-A, BTH-3 displayed an attractive photocatalytic HER of 15.1 mmol h^{-1} g^{-1} in the presence of co-catalyst (Pt) and sacrificial electron donor (0.1 M ascorbic acid).

2.2 Photocatalytic CO_2 reduction

Energy crisis and environmental concerns like greenhouse gas emissions can be addressed by converting CO_2 into chemical fuels using solar energy [76]. In comparison to photocatalytic hydrogen evolution, CO_2 reduction involves more complicated mechanisms. A variety of products could be generated, including carbon monoxide, formic acid, hydrocarbons, and alcohols. As part of the photocatalytic CO_2 reduction system, COFs often served as photosensitive supports [77–82].

A post-synthetic strategy was used by Huang and his colleagues to prepare a re-doped COF for CO_2 reduction (**Figure 9a**) [83]. A triazine-based COF was firstly prepared via reaction between 2,2-bipyridyl-5,5-dialdehyde (BPDA) and 4,4′,4″-(1,3,5-triazine-2,4,6-triyl) trianiline (TTA). After bipyridine ligand was reacted with $Re(CO)_5Cl$, the Re moiety was integrated into COF as a homogeneous photocatalyst for CO_2 reduction. Re-COF was found to be an effective photocatalyst for reducing CO2 to CO with high selectivity (98%) (**Figure 9b**). Because the COFs

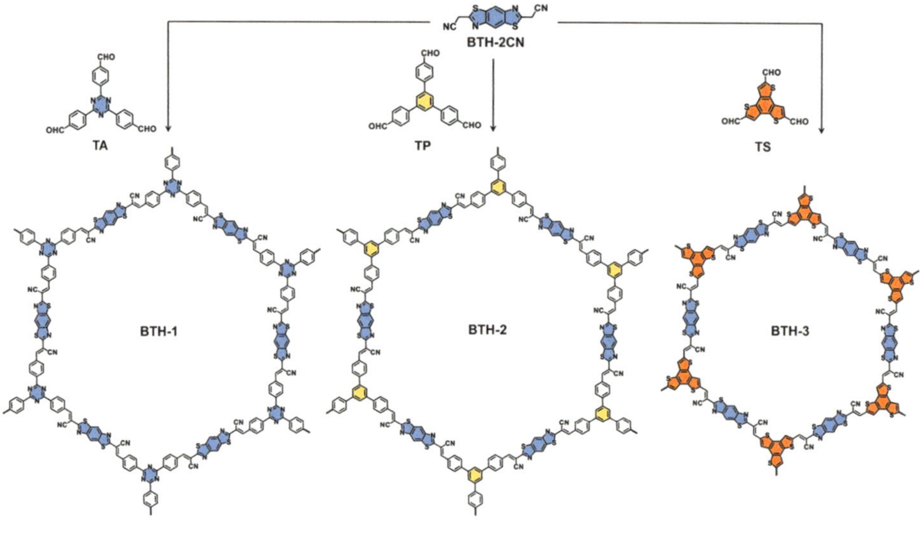

Figure 8.
Synthetic routes and chemical structures of COFs BTH-1, 2, 3.

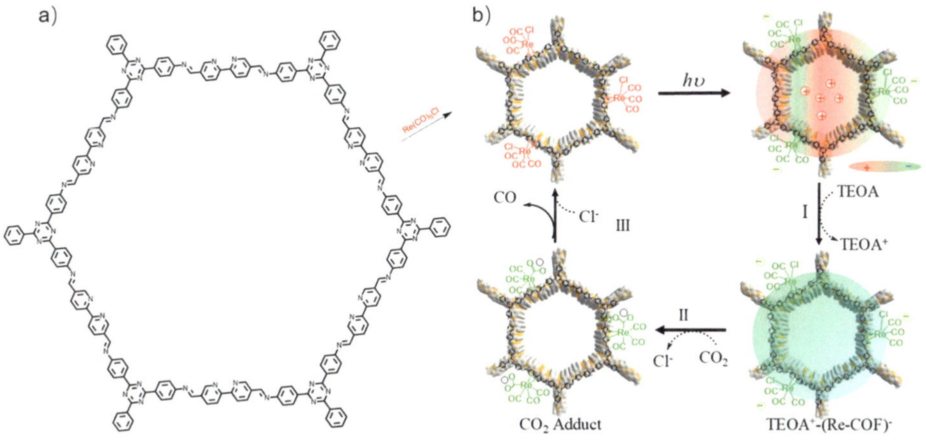

Figure 9.
(a) Synthesis of COF and Re-COF. (b) Proposed Catalytic Mechanism for CO_2 reduction.

are crystalline and porous, they demonstrated better activity than homogeneous Re catalysts. Cooper and co-workers combined the rhenium complex with vinyl-linked conjugated Bpy-sp^{2}c-COF affording a Re-Bpy-sp^{2}c-COF heterogeneous photocatalyst [84]. In this case, bipyridine unit offered the ligation to Re complex, was integrated into the vinyl-linked Bpy-sp^{2}c-COF via Knoevenagel condensation. To produce the desired photocatalyst, [Re(CO)$_5$Cl] was loaded on the Bpy-sp^{2}c-COF. In the presence of TEOA, the yield of CO was 1040 μmol g^{-1} h^{-1} and the selectivity for H_2 was 81% under visible light irradiation. The dye-sensitization process improved selectivity to 86% and CO production rate to1400 μmol g^{-1} h^{-1}. Furthermore, platinum will generate syngas when added. Moreover, platinum could be tuned to adjust the chemical composition of the gas obtained.

A synergistic catalyst (Ni-TpBpy) was developed by Zou and co-workers by combining COF (TpBpy) with a single Ni site. [85]. They first constructed a 2,2′-bipyridine-based COF using 5′-diamino-2,2′-bipyridine. Then, the single Ni sites were loaded via treatment with $Ni(ClO_4)_2$. In the presence of $[Ru(BPY)_3]Cl_2$, the CO yield reached 4057 μmol g^{-1} within 5 hours, and the selectivity was as high as 96%. A unique microenvironment around single Ni sites in TpBpy was found to be responsible for the excellent photoactivity.

Metalloporphyrin (TAPP) complexes have shown good absorption of visible light and potential to reduce CO_2 emissions [86]. Taking that into account, Lan and co-workers constructed a variety of porphyrin-tetrathiafulvalene covalent organic frameworks for CO_2 reduction (**Figure 10**) [87]. By introducing electron-rich tetrathiafulvalene (TTF) molecules into COFs, a donor-acceptor effect was created. Consequently, the generated charge carriers could be efficiently separated and transferred from TTF to TAPP moiety. With TTCOF-Zn as photocatalyst, the yield of CO is 12.33 μmol and the selectivity is close to 100% after 60 hours of visible light irradiation without adding any sacrificant and noble metal cocatalyst. In order to further enhance reduction efficiency, the same group prepared a series of Z-scheme photocatalysts based on COF semiconductors, which combined inorganic semiconductors (TiO_2, Bi_2WO_6, and α-Fe_2O_3) with COFs (COF-316/318) [88]. It was shown that photocatalytic CO production could reach 69.67 μmol g^{-1}h^{-1} without additional photosensitizers and sacrificial agents.

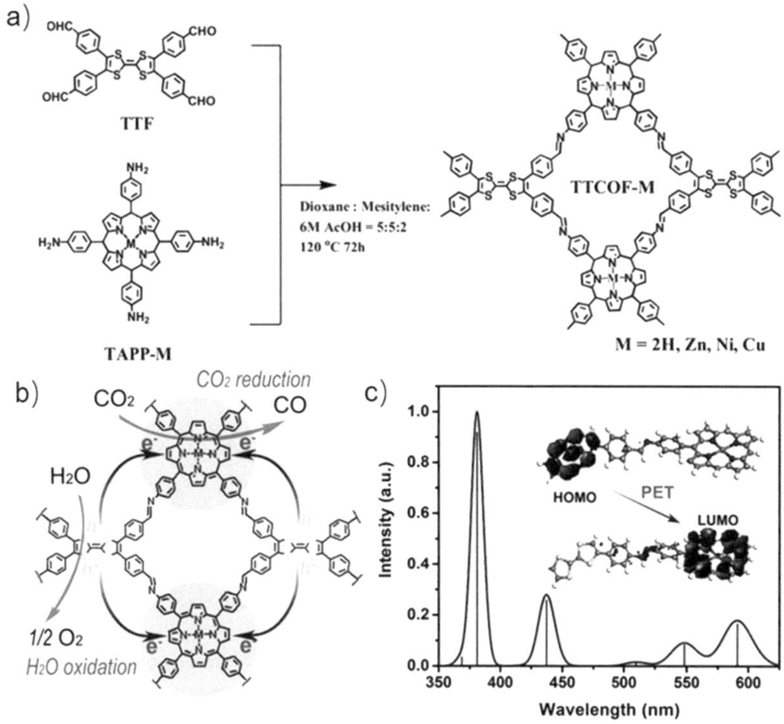

Figure 10.
(a) Schematic of the synthesis of TTCOF-M. (b)Schematic of the mechanism of TTCOF-M CO_2RR. (c)Theoretical simulation UV/Vis DRS of TTCOF-Zn and scheme of PET route under light excitation (inset).

Recent studies have shown that COFs without metal active centers can also be used for photocatalytic reduction of carbon dioxide. It is obvious that the reduction efficiency is much less than that of metallic-supplemented COFs, but metal-free COFs are also of great significance for the development of future carbon dioxide reduction. With these concerns, Liu and co-workers prepared a β-ketoenamine-based 2D COF, termed TpBb-COF (**Figure 11**), by condensation of 2,6-diaminobenzo[1,2-d:4,5-d′] bisthiazole and 1,3,5-triformylphloroglucinol, exhibited an excellent CO production in gas–solid system without using any photosensitizer and sacrificial agent [89]. Interestingly, the reduction of CO_2 concentration was more beneficial to CO production, and when CO_2 concentration was reduced to 30.0%, the CO generation rate increased to 89.9 μmol g^{-1} h^{-1}. The mechanism of photocatalytic reduction of carbon dioxide and the rate equation between the production rate of CO and the concentrations of CO_2 are given.

In recent years, 3D COFs have also made progress in the field of photocatalytic CO_2 reduction. Zhang and co-workers designed eight-linked structural units (TTEP) based on functional porphyrin rings and obtained a 3D COF with pcb topology through Scheff base reaction under solvothermal synthesis conditions (NUST-5 and NUST-6) (**Figure 12**) [90]. Based on the excellent photoelectric characteristics of porphyrin units, these two COFs have shown great application value in the field of CO_2 adsorption and photocatalytic CO_2 reduction. After 10 hours of visible light irradiation, the photocatalytic CO_2 reduction performance test showed that the CO yield of NUST-5 and NUST-6 was 54.7 μmol g^{-1} and

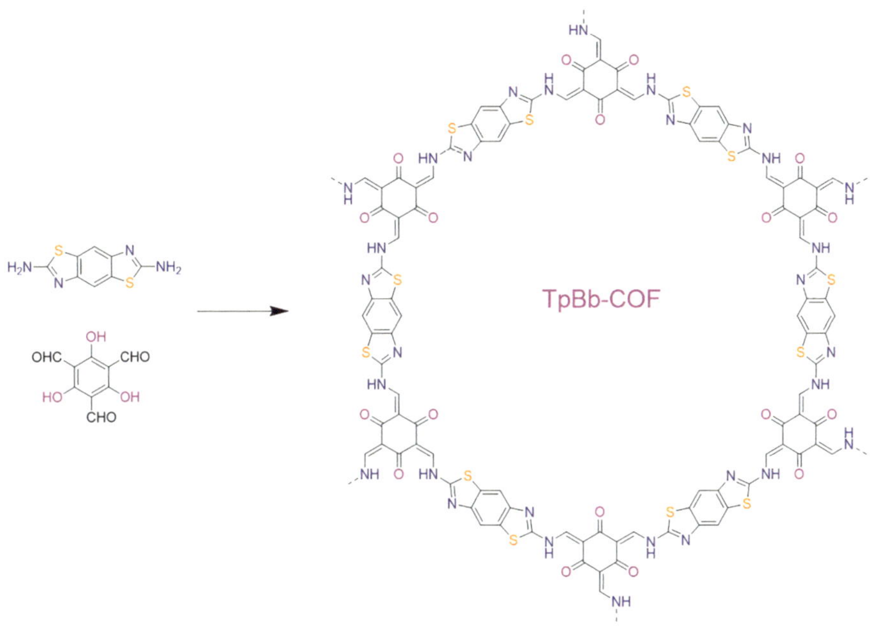

Figure 11.
Schematic of the synthesis of TpBb-COF.

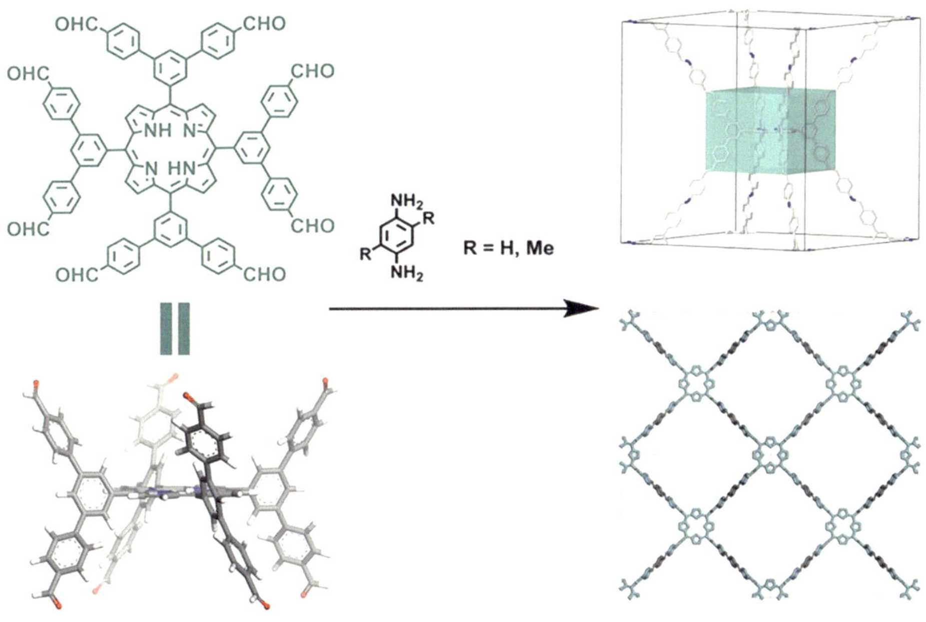

Figure 12.
Illustration for the reticular design and synthesis for NUST-5 (R = H) and NUST-6 (R = Me) 3D COFs with pcb topologies.

76.2 μmol g^{-1}, respectively. In addition, the CH_4 yields of NUST-5 and NUST-6 were 17.2 and 12.8 μmol g^{-1}, respectively. The calculated CO/CH_4 ratios are 76% and 86%.

2.3 Photocatalytic NADH regeneration

As an environmentally friendly and sustainable method of converting solar energy, artificial photosynthesis is modeled after natural photosynthesis. An important component of artificial photosynthesis is coenzyme regeneration (NADH/NADPH) [91]. Many enzyme-mediated synthetic processes require NADH regeneration. Organic polymers with excellent visible light absorption have exhibited excellent photocatalytic performance under the irradiation of visible light [92–94].

A fully sp2-carbon conjugated COF (TP-COF) was synthesized and characterized for use in photocatalytic NADH regeneration in a previous study by Zhao et al. [95]. TP-COF (**Figure 13**) was used as a photosensitizer, while $[Cp^*Rh(bpy)(H)]^+$ was selected as the electron mediator. Within 10 minutes of irradiation with 420 nm light, the NADH regeneration yield reaches 90.4%. The NADH regeneration system was then coupled with L-glutamate dehydrogenase (GDH) (a redox enzyme) to test the activity of the regenerated NADH. Benefiting from the high efficiency of NADH regeneration, α-ketoglutarate was efficiently converted to L-glutamate with a yield of 97% within 12 min.

Also, the linkage effect was investigated in the photocatalytic NADH regeneration [96]. Post-synthesis conversion of two imine-linked COFs (B-COF-1 and T-COF-1)

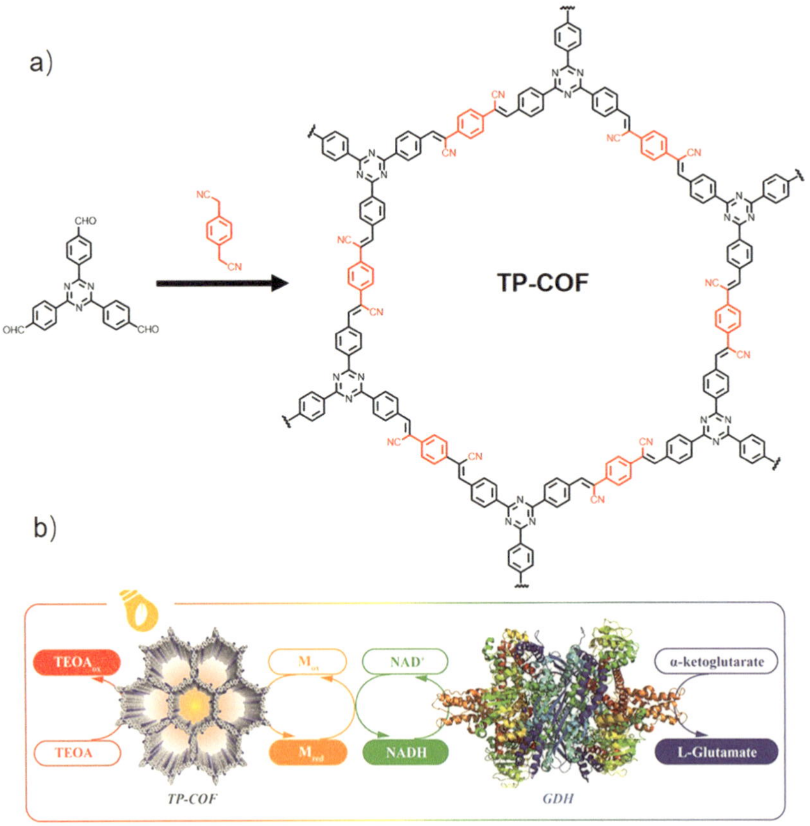

Figure 13.
(a) Synthesis and structures of the TP-COF. (b) Illustration of the artificial PSI-induced coenzyme regeneration and photoenzymatic synthesis of l-glutamate by l-glutamate dehydrogenase (GDH).

into conjugated COFs (B-COF-2 and T-COF-2) was conducted (**Figure 14**). As a result of the subtle structure changes, the photoactivity was entirely different. NADH regeneration yielded 74.0% in 10 min with triazine-containing T-COF-2, much higher than with imine-linked precursors. Subsequently, Chen and co-workers introduced bipyridinium units into the framework structure and constructed a mesoporous alkene-linked COF as a porous solid carrier for co-immobilized formate dehydrogenase (FDH) and Rh-group electron medium [97]. By adjusting the incorporation amount of Rh electron medium, it was beneficial to regenerate NADH from NAD^+, with an apparent quantum yield (AQY) of 9.17 ± 0.44%. Finally, the assembled photocatalyst-enzyme coupling system can selectively convert CO_2 to formic acid, with high efficiency and good reuse.

2.4 Photocatalysis for other organic reactions

The photocatalytic organic reaction has been recognized as a green method for the synthesis of small molecules. In recent years, COFs have been frequently used for photocatalytic synthesis [48, 55, 98]. For instance, Wang and his group reported three COFs (LZU-190, LZU-191, LZU-192) with benzoxazole as the bonding type via a "killing two birds with one stone" strategy (**Figure 15**) [99]. The COFs structure was kept stable in strong acid and base (9 M NaOH, 9 M HCl), trifluoroacetic acid, boiling water, and light (all for 3 days). Subsequently, it was applied in the experiment of photocatalytic oxidation of arylboronic acids to phenols. Under visible light, all three COFs could oxidize arylboronic acids to phenols using air as oxygen source. Subsequently, LZU-190 was taken as an example to study substrate expansion and

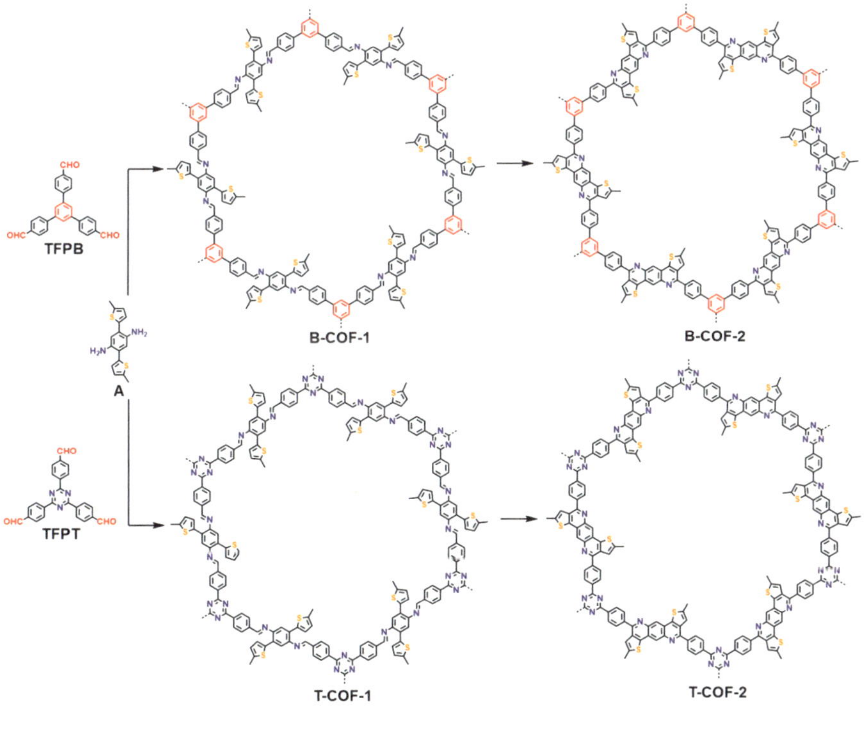

Figure 14.
Synthesis and Structures of B-COF-1,2; T-COF-1,2.

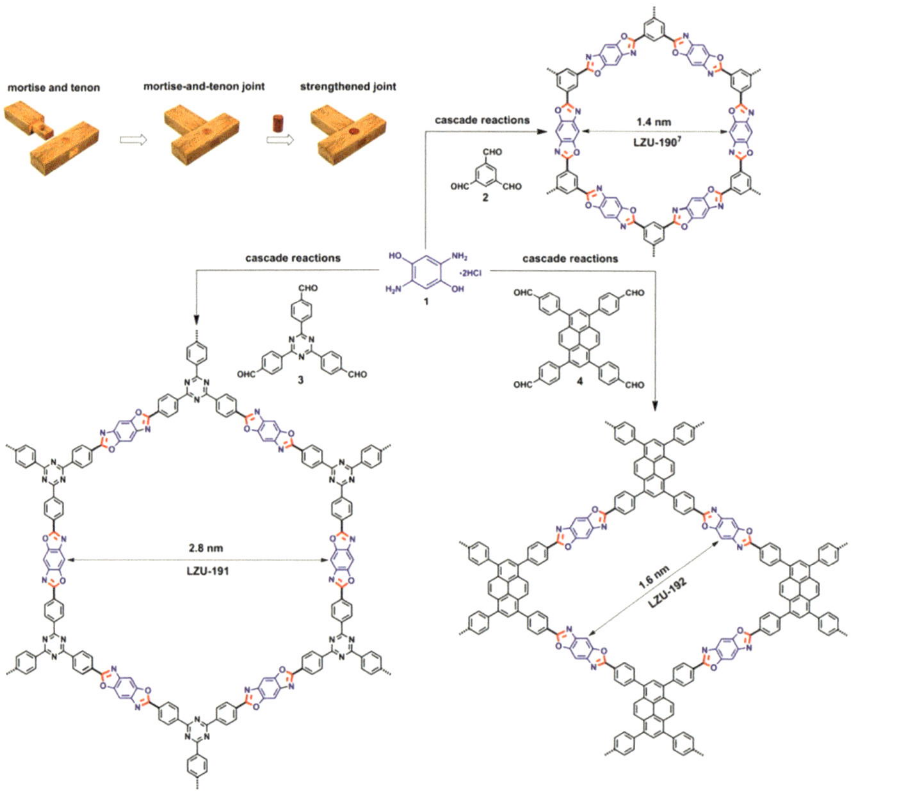

Figure 15.
One-Pot Construction of Benzoxazole-Linked COFs (LZU-190, LZU-191, and LZU-192) via the Cascade Reactions.

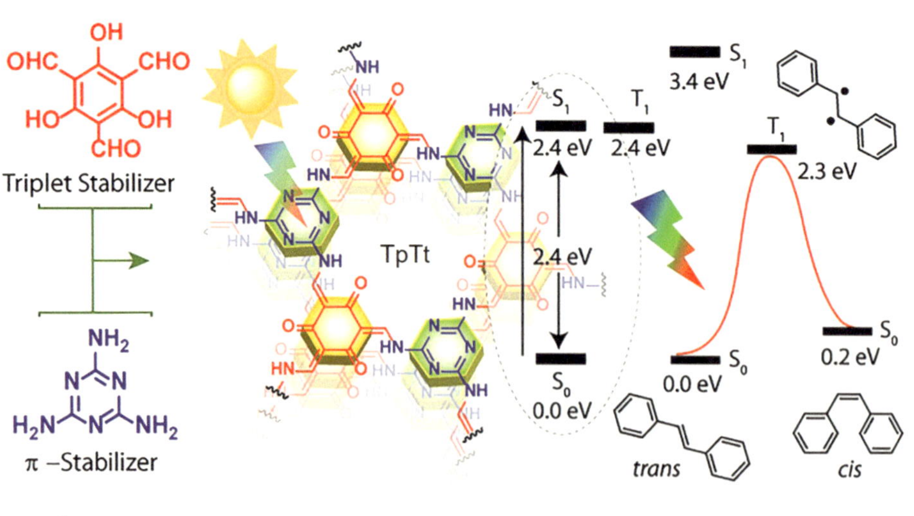

Figure 16.
Schematic representation of the TpTt synthesis and mechanistic representation of trans to cis photoisomerization of stilbene using the TpTt COF catalyst.

mechanism, and electron spin-resonance spectroscopy (ESR) and isotope labeling experiments were used to confirm the mechanism of single electron transfer process.

Banerjee and co-workers designed and synthesized a stable COF(TpTt) based on triazine units for photocatalytic E-Z isomerization of olefins (**Figure 16**) [100]. TpTt has obvious absorption of visible light and can catalyze conversion of trans-stilbene to cis-stilbene under the irradiation of blue light diode. As a result of the good stability of β-ketoenamine linkage, the photoactivity was retained after four cycles. This work also investigated the importance of visible light for this conversion, which begins when light is illuminated and stops when light is closed, suggesting that the reaction takes place through a photocatalytic pathway. In order to further explore the mechanism of photocatalysis, the author uses free radical quenching agent TEMPO for the controllable photocatalytic reaction. When using 4 equiv. TEMPO, cis product yield was significantly reduced to 3%, which confirmed the existence of free radicals and the importance of light catalytic reaction. For different trans olefin substrates, TpTt showed good isomerization of light yield. Besides, Wang et al. developed a red light-driven catalysis system combining the Por-sp^2c-COF with TEMPO [101]. When irradiated with 623 nm red light-emitting diode, amines were converted into imines in minutes using this catalytic system.

3. Conclusions

In summary, COFs have been viewed as promising materials for photocatalysis applications. As a result of the large specific surface area and the porous structure, light and reactants could be absorbed more efficiently, further optimizing photocatalysis. Conjugated structures with a high degree of order could promote direct charge transport and efficient charge transfer. As an added benefit, COF materials are easily synthesized using the versatile toolbox of organic synthesis. The photoactive building blocks can be altered according to practical applications to synthesize COFs with different structures. A variety of influencing factors, such as pore size, band gap, and other factors, can be adjusted based on the structures. The aforementioned characteristics make COFs a suitable photocatalytic platform. Benefiting from these advantages, COFs have shown excellent capacity in harvesting visible light efficiently and have been frequently used for photocatalytic water splitting, CO_2 reduction, coenzyme regeneration, dye degradation, some organic reaction, etc. There are, however, still some problems that need to be solved. It is still challenging to synthesize COFs materials due to the dichotomy of crystallinity and stability. Additionally, the exfoliation of layered 2D COFs during long-term photocatalysis would influence the performance. The exploration of new linkages and synthetic strategies to increase the stability and maintain the crystallinity of the COFs is necessary both for the development of COFs materials and the application of the photocatalyst. Both of these two factors are important to photocatalytic performance. Notably, the application of 3D COFs for photocatalysis is still rare. 3D COFs could avoid the exfoliation problem usually associated with 2D COFs. In addition, the large-scale synthesis of COFs materials is crucial to this field's future. From the point of view of photocatalysis, the efficiency is still relatively low in such as water splitting and CO_2 reduction compared to the traditional inorganic materials. These organic materials should be explored in more detail for some specific applications. Besides, noble metals as co-catalyst and sacrificial agents are normally necessary at the current study lever. It is desirable to avoid the use of noble metals as co-catalyst and sacrificial agents. This could be realized through the structure regulation of the COFs. For example, we could introduce different heteroatom doping (F, Cl, N) and D-A heterojunction structure into structure, which can make the COFs have more reactive active sites and stronger exciton separation and charge transport ability of the COFs material. Furthermore, COFs materials possess highly tunable structures, providing a great advantage for studying photocatalysis mechanisms. By experimenting with different COF catalysts, it is possible to conclude that the structure–activity relationship. We are likely to gain a better understanding of some key points concerning photo-induced charge carriers, such as the separation efficiency and the carrier transport, as time goes on. While traditional photocatalysis has been studied for half a century, the role of COFs materials in photocatalysis is still relatively new. Despite many challenges, we believe COFs materials will offer new development potential in the field of photocatalysis.

Acknowledgements

This work was supported by the National Natural Science Foundation of China (31202117), Natural Science Foundation of Shandong Province (ZR2020ZD38).

Conflict of interest

The authors declare no conflict of interest.

Author details

Hui Liu and Yingjie Zhao*
College of Polymer Science and Engineering, Qingdao University of Science and Technology, Qingdao, China

*Address all correspondence to: yz@qust.edu.cn

References

[1] Fouquet R. Path dependence in energy systems and economic development. Nature Energy. 2016;**1**:16098

[2] Gurney KR, Mendoza DL, Zhou YY, Fischer ML, Miller CC, Geethakumar S, et al. High resolution fossil fuel combustion CO_2 emission fluxes for the United States. Environmental Science & Technology. 2009;**43**:5535-5541

[3] Zhang YY, Yuan S, Feng X, Li HW, Zhou JW, Wang B. Preparation of nanofibrous metal-organic framework filters for efficient air pollution control. Journal of the American Chemical Society. 2016;**138**:5785-5788

[4] Fujishima A, Honda K. Electrochemical photolysis of water at a semiconductor electrode. Nature. 1972;**238**:37-38

[5] Li W, Elzatahry A, Aldhayan D, Zhao DY. Core-shell structured titanium dioxide nanomaterials for solar energy utilization. Chemical Society Reviews. 2018;**47**:8203-8237

[6] Schneider J, Matsuoka M, Takeuchi M, Zhang JL, Horiuchi Y, Anpo M, et al. Understanding TiO_2 photocatalysis: Mechanisms and materials. Chemical Reviews. 2014;**114**:9919-9986

[7] Li Q, Li X, Wageh S, Al-Ghamdi AA, Yu JG. CdS/graphene nanocomposite photocatalysts. Advanced Energy Materials. 2015;**5**:1500010

[8] Low JX, Dai BZ, Tong T, Jiang CJ, Yu JG. In situ irradiated X ray photoelectron spectroscopy investigation on a direct Z-scheme TiO_2/CdS composite film photocatalyst. Advanced Materials. 2019;**31**:1802981

[9] Ong CB, Ng LY, Mohammad AW. A review of ZnO nanoparticles as solar photocatalysts: Synthesis, mechanisms and applications. Renewable & Sustainable Energy Reviews. 2018;**81**:536-551

[10] Martin DJ, Liu GG, Moniz SJA, Bi YP, Beale AM, Ye JH, et al. Efficient visible driven photocatalyst, silver phosphate: Performance, understanding and perspective. Chemical Society Reviews. 2015;**44**:7808-7828

[11] Sachs M, Sprick RS, Pearce D, Hillman SAJ, Monti A, Guilbert AAY, et al. Understanding structure-activity relationships in linear polymer photocatalysts for hydrogen evolution. Nature. Communications. 2018;**9**:4968

[12] Shu G, Li YD, Wang Z, Jiang JX, Wang F. Poly(dibenzothiophene-S,S-dioxide) with visible light-induced hydrogen evolution rate up to 44.2 mmol h^{-1} g^{-1} promoted by K_2HPO_4. Applied Catalysis B-Environmental. 2020;**261**:118230

[13] Sprick RS, Wilbraham L, Bai Y, Guiglion P, Monti A, Clowes R, et al. Nitrogen containing linear poly(phenylene) derivatives for photo-catalytic hydrogen evolution from water. Chemistry of Materials. 2018;**30**:5733-5742

[14] Tseng PJ, Chang CL, Chan YH, Ting LY, Chen PY, Liao CH, et al. Design and synthesis of Cycloplatinated polymer dots as photocatalysts for visible-light-driven hydrogen evolution. ACS Catalysis. 2018;**8**:7766-7772

[15] Liu J, Liu Y, Liu NY, Han YZ, Zhang X, Huang H, et al. Metal-free efficient photocatalyst for stable visible water splitting via a two-electron pathway. Science. 2015;**347**:970-974

[16] Wang G, Zhang T, Yu W, Sun Z, Nie X, Si R, et al. Efficient electronic modulation of g-C_3N_4 photocatalyst by implanting atomically dispersed Ag_1-N_3 for extremely high hydrogen evolution rates. CCS Chemistry. 2022;**4**:2793-2805

[17] Wang XC, Maeda K, Thomas A, Takanabe K, Xin G, Carlsson JM, et al. A metal-free polymeric photocatalyst for hydrogen production from water under visible light. Nature Materials. 2009;**8**:76-80

[18] Han CZ, Dong PH, Tang HR, Zheng PY, Zhang C, Wang F, et al. Realizing high hydrogen evolution activity under visible light using narrow band gap organic photocatalysts. Chemical Science. 2021;**12**:1796-1802

[19] Lee JSM, Cooper AI. Advances in conjugated microporous polymers. Chemical Reviews. 2020;**120**:2171-2214

[20] Li LW, Cai ZX, Wu QH, Lo WY, Zhang N, Chen LX, et al. Rational design of porous conjugated polymers and roles of residual palladium for photocatalytic hydrogen production. Journal of the American Chemical Society. 2016;**138**:7681-7686

[21] Sprick RS, Jiang JX, Bonillo B, Ren SJ, Ratvijitvech T, Guiglion P, et al. Tunable organic photocatalysts for visible-light-driven hydrogen evolution. Journal of the American Chemical Society. 2015;**137**:3265-3270

[22] Yang C, Ma BC, Zhang LZ, Lin S, Ghasimi S, Landfester K, et al. Molecular engineering of conjugated Polybenzothiadiazoles for enhanced hydrogen production by photosynthesis. Angewandte Chemie-International Edition. 2016;**55**:9202-9206

[23] Zeng QR, Cheng ZH, Yang C, He Y, Meng N, Faul CFJ, et al. Metal-free synthesis of pyridyl conjugated microporous polymers for photocatalytic hydrogen evolution. Chinese Journal of Polymer Science. 2021;**39**:1004-1012

[24] Ren X, Li C, Kang W, Li H, Ta N, Ye S, et al. Enormous promotion of photocatalytic activity through the use of near-single layer covalent organic frameworks. CCS Chemistry. 2022;**4**:2429-2439

[25] Stegbauer L, Schwinghammer K, Lotsch BV. A hydrazone-based covalent organic framework for photocatalytic hydrogen production. Chemical Science. 2014;**5**:2789-2793

[26] Wang J, Zhang J, Peh SB, Liu GL, Kundu T, Dong JQ, et al. Cobalt-containing covalent organic frameworks for visible light-driven hydrogen evolution. Science China-Chemistry. 2020;**63**:192-197

[27] Wang XY, Chen LJ, Chong SY, Little MA, Wu YZ, Zhu WH, et al. Sulfone-containing covalent organic frameworks for photocatalytic hydrogen evolution from water. Nature Chemistry. 2018;**10**:1180-1189

[28] Cheng G, Wang KW, Wang SY, Guo LP, Wang ZJ, Jiang JX, et al. Pyrene-based covalent triazine framework towards high-performance sensing and photocatalysis applications. Science China Materials. 2021;**64**:149-157

[29] Liu MY, Jiang K, Ding X, Wang SL, Zhang CX, Liu J, et al. Controlling monomer feeding rate to achieve highly crystalline covalent triazine frameworks. Advanced Materials. 2019;**31**:1807865

[30] Wang KW, Yang LM, Wang X, Guo LP, Cheng G, Zhang C, et al. Covalent triazine frameworks via a Low-temperature polycondensation approach.

Angewandte Chemie-International Edition. 2017;**56**:14149-14153

[31] Wang N, Cheng G, Guo LP, Tan BE, Jin SB. Hollow covalent triazine frameworks with variable Shell thickness and morphology. Advanced Functional Materials. 2019;**29**:1904781

[32] Huang X, Sun C, Feng X. Crystallinity and stability of covalent organic frameworks. Science China-Chemistry. 2020;**63**:1367-1390

[33] Fan HW, Mundstock A, Feldhoff A, Knebel A, Gu JH, Meng H, et al. Covalent organic framework-covalent organic framework bilayer membranes for highly selective gas separation. Journal of the American Chemical Society. 2018;**140**:10094-10098

[34] Gao Q, Li X, Ning GH, Xu HS, Liu CB, Tian BB, et al. Covalent organic framework with frustrated bonding network for enhanced carbon dioxide storage. Chemistry of Materials. 2018;**30**:1762-1768

[35] Kang ZX, Peng YW, Qian YH, Yuan DQ, Addicoat MA, Heine T, et al. Mixed matrix membranes (MMMs) comprising exfoliated 2D covalent organic frameworks (COFs) for efficient CO_2 separation. Chemistry of Materials. 2016;**28**:1277-1285

[36] Li ZP, Zhi YF, Feng X, Ding XS, Zou YC, Liu XM, et al. An azine-linked covalent organic framework: Synthesis, characterization and efficient gas storage. Chemistry-a European Journal. 2015;**21**:12079-12084

[37] Deng YW, Zou T, Tao XF, Semetey V, Trepout S, Marco S, et al. Poly(epsilon-caprolactone)-block-polysarcosine by ring-opening polymerization of sarcosine N-Thiocarboxyanhydride: Synthesis and Thermoresponsive self-assembly. Biomacromolecules. 2015;**16**:3265-3274

[38] Liu H, Yan XL, Chen WB, Xie Z, Li S, Chen WH, et al. Donor-acceptor 2D covalent organic frameworks for efficient heterogeneous photocatalytic alpha-oxyamination. Science China-Chemistry. 2021;**64**:827-833

[39] Liu Y, Wu C, Sun Q, Hu F, Pan Q, Sun J, et al. Spirobifluorene-based three-dimensional covalent organic frameworks with rigid topological channels as efficient heterogeneous catalyst. CCS Chemistry. 2021;**3**:2418-2427

[40] Keller N, Bein T. Optoelectronic processes in covalent organic frameworks. Chemical Society Reviews. 2021;**50**:1813-1845

[41] Ding SY, Dong M, Wang YW, Chen YT, Wang HZ, Su CY, et al. Thioether-based fluorescent covalent organic framework for selective detection and facile removal of mercury(II). Journal of the American Chemical Society. 2016;**138**:3031-3037

[42] Liu XG, Huang DL, Lai C, Zeng GM, Qin L, Wang H, et al. Recent advances in covalent organic frameworks (COFs) as a smart sensing material. Chemical Society Reviews. 2019;**48**:5266-5302

[43] Wu XW, Han X, Xu QS, Liu YH, Yuan C, Yang S, et al. Chiral BINOL-based covalent organic frameworks for enantioselective sensing. Journal of the American Chemical Society. 2019;**141**:7081-7089

[44] Lei ZD, Yang QS, Xu Y, Guo SY, Sun WW, Liu H, et al. Boosting lithium storage in covalent organic framework via activation of 14-electron redox chemistry. Nature. Communications. 2018;**9**:576

[45] Vitaku E, Gannett CN, Carpenter KL, Shen LX, Abruna HD,

Dichtel WR. Phenazine-based covalent organic framework cathode materials with high energy and power densities. Journal of he American Chemical Society. 2020;**142**:16-20

[46] Diercks CS, Yaghi OM. The atom, the molecule, and the covalent organic framework. Science. 2017;**355**:6328

[47] Ding SY, Wang W. Covalent organic frameworks (COFs): From design to applications. Chemical Society Reviews. 2013;**42**:548-568

[48] Geng KY, He T, Liu RY, Dalapati S, Tan KT, Li ZP, et al. Covalent organic frameworks: Design, synthesis, and functions. Chemical Reviews. 2020;**120**:8814-8933

[49] Gui B, Lin GQ, Ding HM, Gao C, Mal A, Wang C. Three-dimensional covalent organic frameworks: From topology design to applications. Accounts of Chemical Research. 2020;**53**:2225-2234

[50] Hu XL, Li HG, Tan BE. COFs-based porous materials for photocatalytic applications. Chinese Journal of Polymer Science. 2020;**38**:673-684

[51] Huang N, Wang P, Jiang DL. Covalent organic frameworks: A materials platform for structural and functional designs. Nature Reviews Materials. 2016;**1**:16068

[52] Kandambeth S, Dey K, Banerjee R. Covalent organic frameworks: Chemistry beyond the structure. Journal of the American Chemical Society. 2019;**141**:1807-1822

[53] Tao S, Jiang D. Covalent organic frameworks for energy conversions: Current status, challenges, and perspectives. CCS Chemistry. 2021;**3**:2003-2024

[54] Waller PJ, Gandara F, Yaghi OM. Chemistry of covalent organic frameworks. Accounts of Chemical Research. 2015;**48**:3053-3063

[55] Wang H, Wang H, Wang ZW, Tang L, Zeng GM, Xu P, et al. Covalent organic framework photocatalysts: Structures and applications. Chemical Society Reviews. 2020;**49**:4135-4165

[56] Banerjee T, Gottschling K, Savasci G, Ochsenfeld C, Lotsch BV. H-2 evolution with covalent organic framework photocatalysts. Acs Energy Letters. 2018;**3**:400-409

[57] Wan YY, Wang L, Xu HX, Wu XJ, Yang JL. A simple molecular design strategy for two-dimensional covalent organic framework capable of visible-light-driven water splitting. Journal of the American Chemical Society. 2020;**142**:4508-4516

[58] Yang Q, Luo ML, Liu KW, Cao HM, Yan HJ. Covalent organic frameworks for photocatalytic applications. Applied Catalysis B-Environmental. 2020;**276**:119174

[59] Schwinghammer K, Tuffy B, Mesch MB, Wirnhier E, Martineau C, Taulelle F, et al. Triazine-based carbon nitrides for visible-light-driven hydrogen evolution. Angewandte Chemie-International Edition. 2013;**52**:2435-2439

[60] Ding YB, Zhu WH, Xie YS. Development of ion Chemosensors based on porphyrin analogues. Chemical Reviews. 2017;**117**:2203-2256

[61] Chen RF, Wang Y, Ma Y, Mal A, Gao XY, Gao L, et al. Rational design of isostructural 2D porphyrin-based covalent organic frameworks for tunable photocatalytic hydrogen evolution. Nature. Communications. 2021;**12**:1354

[62] Yang S, Lv H, Zhong H, Yuan D, Wang X, Wang R. Transformation of covalent organic frameworks from N-Acylhydrazone to oxadiazole linkages for smooth Electron transfer in photocatalysis. Angewandte Chemie-International Edition. 2022;**61**:e202115655

[63] Vyas VS, Haase F, Stegbauer L, Savasci G, Podjaski F, Ochsenfeld C, et al. A tunable azine covalent organic framework platform for visible light-induced hydrogen generation. Nature. Communications. 2015;**6**:8508

[64] Stegbauer L, Zech S, Savasci G, Banerjee T, Podjaski F, Schwinghammer K, et al. Tailor-made photoconductive pyrene-based covalent organic frameworks for visible-light driven hydrogen generation. Advanced Energy Materials. 2018;**8**:1703278

[65] Banerjee T, Haase F, Savasci G, Gottschling K, Ochsenfeld C, Lotsch BV. Single-site photocatalytic H_2 evolution from covalent organic frameworks with molecular Cobaloxime Co-catalysts. Journal of the American Chemical Society. 2017;**139**:16228-16234

[66] Pachfule P, Acharjya A, Roeser J, Langenhahn T, Schwarze M, Schomacker R, et al. Diacetylene functionalized covalent organic framework (COF) for photocatalytic hydrogen generation. Journal of the American Chemical Society. 2018;**140**:1423-1427

[67] Ghosh S, Nakada A, Springer MA, Kawaguchi T, Suzuki K, Kaji H, et al. Identification of prime factors to maximize the photocatalytic hydrogen evolution of covalent organic frameworks. Journal of the American Chemical Society. 2020;**142**:9752-9762

[68] Schwab MG, Hamburger M, Feng X, Shu J, Spiess HW, Wang X, et al. Photocatalytic hydrogen evolution through fully conjugated poly(azomethine) networks. Chemical Communications. 2010;**46**:8932-8934

[69] Li W, Huang X, Zeng T, Liu YA, Hu W, Yang H, et al. Thiazolo[5,4-d]thiazole-based donor-acceptor covalent organic framework for sunlight-driven hydrogen evolution. Angewandte Chemie-International Edition. 2021;**60**:1869-1874

[70] Wang G-B, Li S, Yan C-X, Lin Q-Q, Zhu F-C, Geng Y, et al. A benzothiadiazole-based covalent organic framework for highly efficient visible-light driven hydrogen evolution. Chemical Communications. 2020;**56**:12612-12615

[71] Chen W, Wang L, Mo D, He F, Wen Z, Wu X, et al. Modulating benzothiadiazole-based covalent organic frameworks via halogenation for enhanced photocatalytic water splitting. Angewandte Chemie-International Edition. 2020;**59**:16902-16909

[72] Jin EQ, Lan ZA, Jiang QH, Geng KY, Li GS, Wang XC, et al. 2D sp(2) carbon-conjugated covalent organic frameworks for photocatalytic hydrogen production from water. Chem. 2019;**5**:1632-1647

[73] Bi S, Yang C, Zhang WB, Xu JS, Liu LM, Wu DQ, et al. Two-dimensional semiconducting covalent organic frameworks via condensation at arylmethyl carbon atoms. Nature Communications. 2019;**10**:2467

[74] Wei SC, Zhang F, Zhang WB, Qiang PR, Yu KJ, Fu XB, et al. Semiconducting 2D triazine-cored covalent organic frameworks with unsubstituted olefin linkages. Journal

of the American Chemical Society. 2019;**141**:14272-14279

[75] Wang YC, Hao WB, Liu H, Chen RZ, Pan QY, Li ZB, et al. Facile construction of fully sp^2-carbon conjugated two-dimensional covalent organic frameworks containing benzobisthiazole units. Nature. Communications. 2022;**13**:100

[76] Dhakshinamoorthy A, Navalon S, Corma A, Garcia H. Photocatalytic CO_2 reduction by TiO_2 and related titanium containing solids. Energy & Environmental Science. 2012;**5**:9217-9233

[77] Gong YN, Zhong WH, Li Y, Qiu YZ, Zheng LR, Jiang J, et al. Regulating photocatalysis by spin-state manipulation of cobalt in covalent organic frameworks. Journal of the American Chemical Society. 2020;**142**:16723-16731

[78] Kim YH, Kim N, Seo JM, Jeon JP, Noh HJ, Kweon DH, et al. Benzothiazole-based covalent organic frameworks with different symmetrical combinations for photocatalytic CO_2 conversion. Chemistry of Materials. 2021;**33**:8705-8711

[79] Nguyen HL, Alzamly A. Covalent organic frameworks as emerging platforms for CO_2 photoreduction. ACS Catalysis. 2021;**11**:9809-9824

[80] Skorjanc T, Shetty D, Mahmoud ME, Gandara F, Martinez JI, Mohammed AK, et al. Metallated Isoindigo-porphyrin covalent organic framework photocatalyst with a narrow band gap for efficient CO_2 conversion. ACS Applied Materials & Interfaces. 2022;**14**:2015-2022

[81] Wang XY, Fu ZW, Zheng LR, Zhao C, Wang X, Chong SY, et al. Covalent organic framework nanosheets embedding single cobalt sites for photocatalytic reduction of carbon dioxide. Chemistry of Materials. 2020;**32**:9107-9114

[82] Yang Y, Lu Y, Zhang HY, Wang Y, Tang HL, Sun XJ, et al. Decoration of active sites in covalent-organic framework: An effective strategy of building efficient photocatalysis for CO_2 reduction. ACS Sustainable Chemistry & Engineering. 2021;**9**:13376-13384

[83] Yang SZ, Hu WH, Zhang X, He PL, Pattengale B, Liu CM, et al. 2D covalent organic frameworks as intrinsic photocatalysts for visible light-driven CO_2 reduction. Journal of the American Chemical Society. 2018;**140**:14614-14618

[84] Fu ZW, Wang XY, Gardner A, Wang X, Chong SY, Neri G, et al. A stable covalent organic framework for photocatalytic carbon dioxide reduction. Chemical Science. 2020;**11**:543-550

[85] Zhong WF, Sa RJ, Li LY, He YJ, Li LY, Bi JH, et al. A covalent organic framework bearing single Ni sites as a synergistic photocatalyst for selective photoreduction of CO_2 to CO. Journal of the American Chemical Society. 2019;**141**:7615-7621

[86] Kilsa K, Kajanus J, Macpherson AN, Martensson J, Albinsson B. Bridge-dependent electron transfer in porphyrin-based donor-bridge-acceptor systems. Journal of the American Chemical Society. 2001;**123**:3069-3080

[87] Lu M, Liu J, Li Q, Zhang M, Liu M, Wang JL, et al. Rational design of crystalline covalent organic frameworks for efficient CO_2 photoreduction with H2O. Angewandte Chemie-International Edition. 2019;**58**:12392-12397

[88] Zhang M, Lu M, Lang ZL, Liu J, Liu M, Chang JN, et al. Semiconductor/covalent-organic-framework Z-scheme heterojunctions for

artificial photosynthesis. Angewandte Chemie-International Edition. 2020;**59**:6500-6506

[89] Cui J-X, Wang L-J, Feng L, Meng B, Zhou Z-Y, Su Z-M, et al. A metal-free covalent organic framework as a photocatalyst for CO_2 reduction at low CO_2 concentration in a gas-solid system. Journal of Materials Chemistry A. 2021;**9**:24895-24902

[90] Shan Z, Wu M, Zhu D, Wu X, Zhang K, Verduzco R, et al. 3D covalent organic frameworks with interpenetrated pcb topology based on 8-connected cubic nodes. Journal of the American Chemical Society. 2022;**144**:5728-5733

[91] Wang XD, Saba T, Yiu HHP, Howe RF, Anderson JA, Shi JF. Cofactor NAD(P)H regeneration inspired by heterogeneous pathways. Chem. 2017;**2**:621-654

[92] Liu J, Huang JH, Zhou H, Antonietti M. Uniform graphitic carbon nitride nanorod for efficient photocatalytic hydrogen evolution and sustained Photoenzymatic catalysis. Acs Applied Materials & Interfaces. 2014;**6**:8434-8440

[93] Pan QY, Liu H, Zhao YJ, Chen SQ, Xue B, Kan XN, et al. Preparation of N-Graphdiyne nanosheets at liquid/liquid Interface for photocatalytic NADH regeneration. ACS Applied Materials & Interfaces. 2019;**11**:2740-2744

[94] Wang YC, Liu H, Pan QY, Ding NX, Yang CM, Zhang ZH, et al. Construction of Thiazolo 5,4-d thiazole-based two-dimensional network for efficient photocatalytic CO2 reduction. Acs Applied Materials & Interfaces. 2020;**12**:46483-46489

[95] Zhao YJ, Liu H, Wu CY, Zhang ZH, Pan QY, Hu F, et al. Fully conjugated two-dimensional sp^2-carbon covalent organic frameworks as artificial photosystem I with high efficiency. Angewandte Chemie-International Edition. 2019;**58**:5376-5381

[96] Wang YC, Liu H, Pan QY, Wu CY, Hao WB, Xu J, et al. Construction of fully conjugated covalent organic frameworks via facile linkage conversion for efficient Photoenzymatic catalysis. Journal of the American Chemical Society. 2020;**142**:5958-5963

[97] Zhao Z, Zheng D, Guo M, Yu J, Zhang S, Zhang Z, et al. Engineering olefin-linked covalent organic frameworks for Photoenzymatic reduction of CO_2. Angewandte Chemie International Edition. 2022;**61**:e202200261

[98] Chen D, Chen WB, Zhang G, Li S, Chen WH, Xing GL, et al. N-rich 2D heptazine covalent organic frameworks as efficient metal-free photocatalysts. ACS Catalysis. 2022;**12**:616-623

[99] Wei PF, Qi MZ, Wang ZP, Ding SY, Yu W, Liu Q, et al. Benzoxazole-linked Ultrastable covalent organic frameworks for photocatalysis. Journal of the American Chemical Society. 2018;**140**:4623-4631

[100] Bhadra M, Kandambeth S, Sahoo MK, Addicoat M, Balaraman E, Banerjee R. Triazine functionalized porous covalent organic framework for photo-organocatalytic E-Z isomerization of olefins. Journal of the American Chemical Society. 2019;**141**:6152-6156

[101] Shi JL, Chen RF, Hao HM, Wang C, Lang XJ. 2D sp^2 carbon-conjugated porphyrin covalent organic framework for cooperative photocatalysis with TEMPO. Angewandte Chemie-International Edition. 2020;**59**:9088-9093

Chapter 5

Applications of Covalent Organic Frameworks (COFs) in Oncotherapy

Guiyang Zhang

Abstract

Covalent organic frameworks (COFs) are emerging organic crystalline polymer materials, which are formed by reversible condensation reactions between lightweight molecular fragments. They have excellent properties such as low density, good porosity and crystallinity, and high thermal stability. These materials are biodegradable due to the reversible condensation process between the monomers. Compared with another widely studied material with metal-organic frameworks, COFs have no additional toxicity caused by introducing metal ions. Therefore, a high potential exists in biomedicine. The chapter aimed to introduce the application of biomaterial COFs in oncotherapy and identify the specific advantages of different types of COFs for specific biomedical applications.

Keywords: covalent organic frameworks, nanomaterials, oncotherapy, drug carriers, combination therapy

1. Introduction

Porous materials have significantly affected biomedical applications and have been widely used in biomaterials, drug delivery, immune engineering, tissue engineering, and biomedical devices [1–3]. Note that these porous materials have the advantage of encapsulating drugs within their pores and can sustain drug release in drug delivery [2, 4]. Porous polymeric materials traditionally used in biomedicine are amorphous, so there is no porosity in the matrix to optimally encapsulate drugs. For example, poly (lactide-co-glycolide) acid (PLGA) has been widely used as a biomaterial for preclinical and clinical research [5–7]. However, polylactic acid-glycolic acid copolymers are usually amorphous and have no well-defined porous structure. Therefore, the synthesis of biomaterials should be optimized to maximize the drug encapsulation efficiency. On the other hand, high-purity crystalline materials have a well-defined porous structure, which significantly affects the biomedical field [8, 9]. Nanocarriers generally have multiply layers, mesoporous, and hollows, and their unique nanometer size, pore structure, and good biocompatibility make them good drug carriers [10]. The advantages of the current application of nanomaterials in tumors mainly include the following aspects: (1) large specific surface area and high drug encapsulation efficiency; (2) improving the permeability of cell membranes and drug utilization, with efficient drug delivery; (3) a long half-life *in vivo*; (4) controllable release in the slightly acidic environment of tumors; (5) the functionalized surface [11]. Ideal cancer therapy cannot be achieved due to the current single treatment method. The use of nanomaterials as carriers for drug co-delivery or the integration of multiple therapeutic modalities can overcome the inherent shortcomings of single therapeutic strategies. Organic frameworks are newly developed crystalline porous composites in recent years. Metal/covalent-organic frameworks have been widely studied and used for oncotherapy due to their advantages such as good biocompatibility, large specific surface area, and easy modification [12].

Since COFs were first reported by Yaghi and his colleagues in 2005, the development of COFs with unique structural features and properties has attracted great research interest in the scientific community [13]. It has great application potential in biomedical applications due to its unique properties, such as controllable pore size, easy modification, high stability, and good biocompatibility [14]. COFs are promising new porous materials synthesized by covalent bonding. Note that the 2D or 3D porous crystal structures with specific spatial subunit organization endow them with highly desirable features: low density, structural diversity, porous structure, high specific surface area, free of heavy metal ions, and tunable pore size [14, 15]. The covalent bonds (B-O and C-N bonds) in the common structures of COFs are chemically and thermally stable. Importantly, the internal ordered structure of COFs is different from other groups of covalent polymers. Organic molecules linked by covalent bonds produce short-range ordered structures within small regions, resulting in amorphous or semi-crystalline materials. Interestingly, slow and reversible reactions between organic ligands or building blocks enable these short-range ordered structures to grow, which generates long-range structures and forms COFs. Besides reversible reactions, other features (rigidity) also contribute to the formation of structurally regular COFs [14, 15]. For structurally regular COFs synthesis, these factors require some basic preconditions, such as the reaction and choice of rigid building blocks for COFs. The pore size and porous structure of COFs are affected by the molecular length and established element type. Based on the above characteristics, reversible reactions have been generally recognized as the preferred synthetic route to generate COFs [16].

Materials are generally subjected to post-modification in exploring material applications to make them have more excellent properties, or achieve the functional effect consistent with the expectation. Functional groups are modified on the material surface by chemical or physical methods, and the same is true for studying COFs. COFs expand their applications in catalysis, sensing, separation, and drug delivery after these functional modifications. The advantages of COFs are mainly reflected in the following aspects: (1) COFs are easy to modify due to their structural characteristics and can realize biomedical applications such as tumor targeting, fluorescence imaging, and cancer therapy. (2) The cavities of COFs can encapsulate guest molecules due to their inherent porosity; therefore, they can serve as carriers for drug delivery. (3) COF monomers have different energy level structures under the framework due to their conjugated structure. (4) COFs do not contain metal ions, which can avoid the biological toxicity caused by metal elements [17, 18]. In addition, the inherent limitations of COFs, such as poor physiological stability, poor dispersion, and non-specific targeting, limit their applications in cancer therapy [19–22].

However, based on their superior properties, COFs are still considered a promising and efficient organic material platform for cancer therapy. The section summarizes the applications of COFs in oncotherapy. Although the preparations and applications of COFs are still in their infancy and face many challenges, the huge potential of COFs is bound to provide some new approaches for future oncotherapy and pharmaceutical fields.

2. Research progress in the application of COFs in the drug therapy of tumors

Recently, COFs as carriers have been widely used in cancer therapy research. COFs can deliver drugs to targets and use their photothermal and photodynamic effects to kill tumor cells. The applications of COFs in oncotherapy mainly include the following four types: chemotherapy, photodynamic therapy, photothermal therapy, and combination therapy.

2.1 Application of COFs in chemotherapy

Chemotherapy is one of the main methods for the clinical treatment of tumors. However, the efficacy of chemotherapeutic drugs is limited by many factors, such as poor stability and dispersibility in aqueous solutions, low permeability of cell membranes, non-specific targeting, and uncontrollable drug release. Therefore, the use of chemotherapy drugs directly in patients has certain limitations. COFs-based porous materials have become ideal materials for drug delivery due to their excellent properties such as large specific surface area, porosity, and tunable pore size [23]. COFs as nanocarriers for drug delivery have the following advantages: (1) The coordination bonds in COFs are reversible, which makes them biodegradable. (2) With high drug loading, drugs can be encapsulated through surface mesopores. However, their dispersibility in water is poor compared with other common porous materials, such as nano-silica and polymers.

Zhao's research group synthesized two-dimensional COFs (PI-2-COF and PI-3-COF) in 2016. They have good biocompatibility-existing stably in water in the form of nanoparticles and maintaining a good pore structure. Three different drugs, 5-fluorouracil (5-FU), captopril, and ibuprofen (IBU), were loaded using these two 2D COFs, respectively. The drug-loaded COFs are characterized by powder X-ray diffraction (PXRD), thermogravimetric analysis (TGA) and Fourier transform infrared spectroscopy (FT-IR), and the drug loading effect of COFs is determined by TGA. The results show that the drug loading is as high as 30%. The survival rate of MCF-7 cells treated by the drug-loading system is significantly reduced [24]; however, the unmodified COFs-based drug-loading system has no cancer-cell targeting.

Banerjee's group selected TpASH to modify COFs and synthesized and prepared folic acid-conjugated covalent organic nanosheets (CONs) for targeting in 2017 (**Figure 1a**). Such targeted CONs deliver the drug 5-FU to breast cancer cells via endocytosis and kill them. Ultraviolet-visible (UV-Vis) absorption spectroscopy analysis showed that such CONs are loaded with 12% 5-FU. Cell migration and cellular uptake are investigated by MTT experiments and fluorescence microscopy, respectively, indicating that the system has good cancer-cell targeting. Although this system has great potential value in targeted drug delivery, its low drug load limits its development to some extent [25].

Lin's group dissolved 1,3,5 Tris(4-aminophenylbenzene) and 2,5 dimethoxyterephthalaldehyde in a solvent to explore a simpler assembly process of COFs and drugs in 2019. The reaction is carried out under the catalysis of acetic acids to obtain TAPB DMTP COF. Then DOX and TAPB-DMTP-COF are assembled into a COFs-based

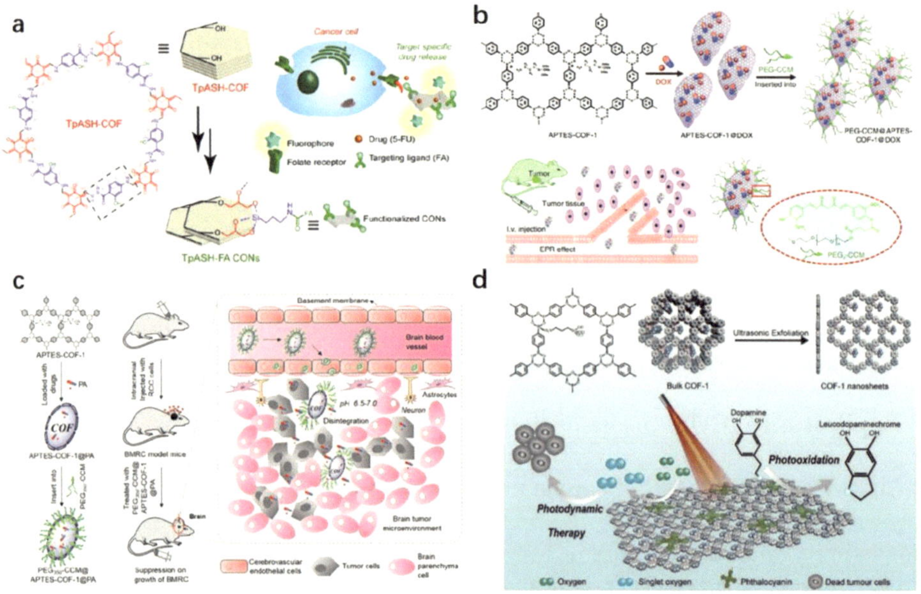

Figure 1.
(a) Structure of TpASH-FA-CONs and cancer therapy [25]. Reproduced with permission from ref. [25]. Copyright 2017, American Chemical Society; (b) Synthesis process and oncotherapy of DOX-loaded PEG-CCM@ APTES-COF-1 nanomaterials [26]. Reproduced with permission from ref. [26]. Copyright 2018, Springer Nature; (c) Synthesis scheme of thin platelet-polymer COFs-based nanocomposite, establishment of the tumor model, and treatment scheme for BMRC [27]. Reproduced with permission from ref. [27]. Copyright 2020, Royal Society of Chemistry; and (d) PcS@COF-1-mediated photooxidation and photodynamic therapy [28]. Reproduced with permission from ref. [28]. Copyright 2020, Royal Society of Chemistry.

drug-loading system. COFs have good binding to DOX by FT-IR results. According to the absorption intensity of DOX in the UV-Vis spectrum, the drug loading of the system is about 32.1 wt%, which is much higher than previously published results. Besides, *in vivo* experiments are performed in mice by intratumoral injection, showing that the system has a good killing effect on cancer cells [29].

In addition to the modification of targeting groups on the surface of COFs, polyethylene glycol (PEG) derivatives can also be modified to enhance their hydrophilicity, prolong their circulation time *in vivo*, and facilitate their accumulation at tumor sites. Jia et al. prepared a series of PEG-modified COF nanomedicines PEG_X-CCM@APTES-COF-1@DOX (X = 350, 1000, and 2000) with better dispersibility (**Figure 1b**). The nano-drug is self-assembled from curcumin (CCM)-modified PEG and amino-functionalized APTES-COF-1@DOX [26]. PEGX-CCM coatings endow the nanomaterials with fluorescence imaging capability and significantly enhance the cellular uptake of materials, the *in vivo* blood circulation time, and the accumulation capability at tumor sites. APTES-COF-1 can be dissociated under acidic conditions to release the internally loaded DOX. *In vitro* fluorescence imaging shows that the PEG_{2000}-CCM@APTES-COF-1@DOX group accumulates more drugs in tumor tissues than in the other groups. *In vivo* anti-tumor experiments prove that PEG_{2000}-CCM@ APTES-COF-1@DOX has an excellent tumor-inhibiting effect compared with other groups.

Subsequently, Zhang et al. developed an amphiphilic platelet-like polymer, COFs-based nanocomposite (PEG_{350}-CCM@APTES-COF-1@PA). Efficient delivery of the tyrosine kinase inhibitor drug PA can be achieved for treating brain metastases from renal cancer (BMRC) (**Figure 1c**) [27]. As a drug delivery vector, the nanomaterials can enhance drug retention in brain tumors. *In vivo* imaging experiments demonstrates that these COFs-based nanocomposites can cross the blood-brain barriers in mice and accumulate at orthotopic intracranial tumor sites of BMRC. The acidic environment of intracranial tumors degrades COF-based polymer nanocomposites due to boronate ester bonding and releases the drugs, which finally inhibits BMRC.

To sum up, chemotherapy drugs still occupy a vital position in cancer treatment. However, traditional drug delivery systems lack targeting, and drugs are distributed everywhere in the body, resulting in various serious toxic and side effects. To solve this problem, COFs have been developed as carriers for traditional chemotherapeutic drugs recently. The drug-loading system enhances the effect of permeability and retention (EPR) at tumor sites through nanocrystallization, which improved the accumulation of the nano-drug system at tumor sites. Meanwhile, COFs are easy to modify, and specific modifications can improve hydrophilicity and targeting as well as reduce the toxic and side effects. Based on their inherent tunable porous structures, the drug loading and drug release rate can be adjusted by adjusting pore sizes and polarity of COFs to meet different drug uses. Most of the reports show that the use of COFs to deliver chemotherapeutic drugs is targeted and controllable, and the efficacy of the compound system is often better than that of single chemotherapeutic drugs. Therefore, other therapies should be combined to exert the synergistic effect between different therapies, which achieves high anti-tumor efficacy and reduce adverse reactions after medication.

2.2 Application of COFs in photodynamic therapy

Similar to chemotherapeutic drug delivery, the improved effect of nanocarriers on photodynamic therapy (PDT) is mainly reflected in the enhanced accumulation of photosensitizer molecules at tumor sites. Traditional photosensitizers are usually organic molecules with a wide range of conjugation systems. They have poor water solubility and tend to aggregate. Small molecules of photosensitizers accumulate less in tumor tissue after systemic administration, which is difficult to satisfy *in vivo* applications. Combining small photosensitizer molecules with nanocarriers can make up for the above deficiencies through passive and active targeting.

On the other hand, nanomaterials can improve the photochemical properties of photosensitizers. The loading of photosensitizers inside nanoparticles can prevent their aggregation at the molecular level, which avoids fluorescence quenching and improves the quantum yield of $^1O_2$. The photosensitizers can be adsorbed on the COF surface to prepare COF-photosensitizer materials using the interaction between COFs and photosensitizers. Yuan et al. used APTES-COF-1 nanosheets to adsorb phthalocyanine photosensitizers to prepare PcS@COF-1 (**Figure 1d**) [28]. Phthalocyanine dyes are highly dispersed on the APTES-COF-1 surface, so PcS@COF-1 exhibits good photodynamic performance under 660-nm laser irradiation, which has obvious killing effects on CT26 cells.

Lin et al. used the COF as a nano template to grow gold nanoparticles (Au NPs) on the COF surface in 2020. Hyaluronic acid (HA) is introduced to improve biocompatibility after covering the material surface with a thin layer of manganese dioxide (MnO_2) (**Figure 2a**) [30]. Synthetic product COF-Au-MnO_2 participates in multiple processes in the tumor microenvironment, which forms a cascade reaction. COF-Au-MnO_2 first reacts with intratumoral H_2O_2 to generate O_2 in the tumor hypoxic environment, which enhances type-II PDT. Next, Au NPs can decompose glucose to

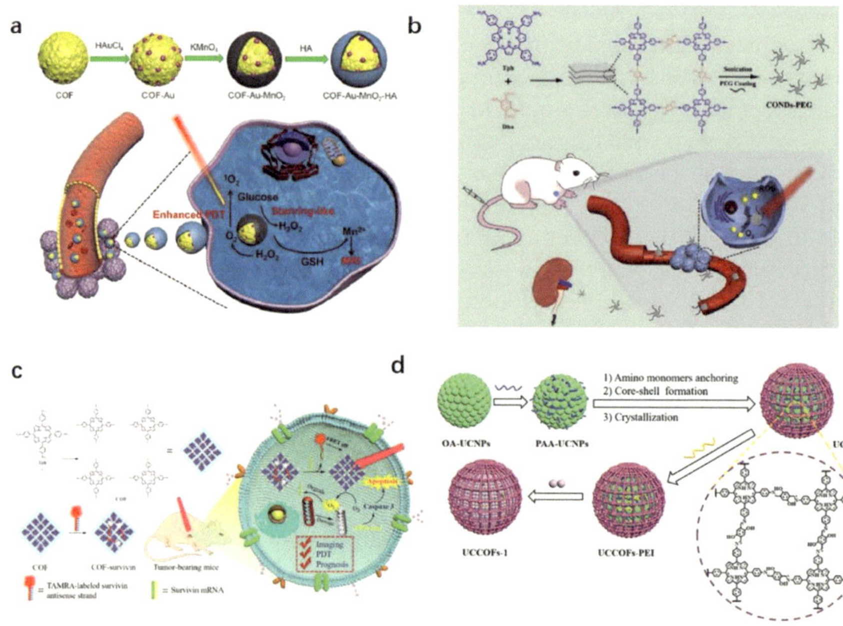

Figure 2.
(a) Synthesis process and therapeutic mechanism of COF-Au-MnO2-HA [30]. Reproduced with permission from ref. [30]. Copyright 2020, Springer Nature; (b) Preparation process of CONDs-PEG nanodots and the use of COF nanoparticles as photodynamic agents for cancer treatment [31]. Reproduced with permission from ref. [31]. Copyright 2019, Elsevier Ltd; (c) Fabrication of therapeutic COF nanoplatforms for tumor imaging, PDT, and prognostic assessment applications [32]. Reproduced with permission from ref. [32]. Copyright 2020, Royal Society of Chemistry; and (d) Synthesis and working of the COFs-based nanoplatform [33]. Reproduced with permission from ref. [33]. Copyright 2019, Royal Society of Chemistry.

generate H_2O_2, which promotes starvation treatment and increases O_2 concentration in tumor tissues. Besides, MnO_2 depletes glutathione (GSH) to enhance the antitumor effect. The released Mn^{2+} is used for T_1-weighted magnetic resonance imaging (MRI). Both *in vitro* and *in vivo* experiments prove that COF-Au-MnO_2 nanoparticles have good tumor-killing effects.

Given the unique structural advantages of COFs, photosensitizers can directly participate in constructing COFs as monomers. The photosensitivity of some porphyrin-based COFs has been reported; however, their applications in PDT are still scarce. Qu et al. synthesized the ultra-small nanodots of porphyrin-based TphDha COFs (**Figure 2b**) with PDT properties [31]. First, TphDha COFs are synthesized in a Pyrex test tube using tetraaldehyde phenyl porphyrin and 2,5-dihydroxyterephthalaldehyde as starting materials. Subsequently, ultrasonic stripping and surface modification with DSPE-PEG are performed. Finally, the CONDs-PEG nanodots are obtained by filtration and separation. The PEG-coated COF nanodots have good physiological stability and biocompatibility. The uniformly dispersed porphyrin molecules on the COF surface endow the CONDs-PEG nanodots with superior light-triggered-induced reactive oxygen species (ROS) generation ability, which shows the excellent PDT effect and good tumor accumulation ability.

COFs have emerged as a promising material for analysis and biomedicine. However, simultaneous use of COFs for cancer diagnosis and treatment remains a challenge. Tang's group reported a COF-based therapeutic nano-platform. Dye-labeled oligonucleotides (TSAS) are modified on porphyrin-based COF nanoparticles (COF NPs) for efficient cancer diagnosis and treatment (**Figure 2c**) [32]. The fluorescence of dyes on TSAS is quenched by COFs *via* fluorescence resonance energy transfer (FRET). When tumor marker mRNA exists, TSAS form more stable double strands and dissociate from COF NPs. Fluorescence signal recovery can be used for selective cancer imaging. Moreover, porphyrin-based COF NPs generate a large amount of ROS through PDT under near-infrared laser irradiation to induce cancer-cell apoptosis. *In vitro* and *in vivo* experiments demonstrate that the COF nano-platform can specifically recognize cancer cells and be used for oncotherapy.

Real-time and in situ monitoring of ROS generation is critical to minimize non-specific damage caused by the high doses of ROS required during PDT. However, phototherapeutic agents generating ROS-related imaging signals during PDT are rare, which prevents easy prediction of future treatment outcomes. Tan *et al*. developed an upconverting COFs-based nano-platform with upconverting nanoparticles (UCNPs) as the core. A layer of TphDha COF was grown in situ on its surface to realize near-infrared (NIR) light-excited PDT (**Figure 2d**) [33]. When the UCNP core is excited with a 980-nm laser, the emissions at 541 and 654 nm are absorbed by the TphDha COF shell to produce $^1O_2$. When the shell thickness was 15 nm, the cell hoes of TphDha COFs have the optimal ability to generate $^1O_2$ as well as the optimal inhibitory effect on HeLa cells. $^1O_2$-labile indocyanine green (ICG) fluorescent dyes are loaded into the pores of TphDha COFs, which enables in situ monitoring of $^1O_2$ and oncotherapy *in vivo*.

Ca^{2+} is a ubiquitous but subtle regulator of cellular physiology tightly controlled within cells. However, intracellular Ca^{2+} regulation, such as mitochondrial Ca^{2+} buffering capacity, may be disrupted by $^1O_2$. Therefore, intracellular Ca^{2+} is overloaded by the synergistic effect of $^1O_2$ and exogenous Ca^{2+} delivery, which is recognized as one of the important pro-death factors. Dong et al. constructed NCOF-based nanoformulation $CaCO_3$@COF-BODIPY-2I@GAG [34]. The small molecules of BODIPY-2I photosensitizer are modified with amino residues on its surface, and $CaCO_3$ NPs are loaded in the COFs-based channel. Finally, glycosaminoglycan (GAG) molecules are surface-modified for targeting CD44 receptors in gastrointestinal tumor cells. After intravenous injection of $CaCO_3$@COF-BODIPY-2I@GAG NPs, the premature leakage of $CaCO_3$ NPs during *in vivo* circulation was avoided under the protection of COFs. $CaCO_3$ NPs are decomposed in the acidic environment of lysosomes to release Ca^{2+} after entry into tumor cells *via* CD44 receptor-mediated endocytosis. Besides, BODIPY-2I covalently attached to the surface can generate 1O2 for PDT under laser irradiation. *In vitro* and *in vivo* experiments demonstrate that the PDT synergistic treatment mode of Ca^{2+} overloads can significantly increase the anti-tumor efficiency.

Photodynamic therapy does not induce tolerance to treatments due to its low toxicity. The therapy is considered to be a promising mode of oncotherapy due to its unique advantages such as low invasiveness for patients. However, the current application of photodynamic therapy is mainly in superficial and flat lesions. The lesions are observed through endoscopy, and it is taken as adjuvant therapy for surgery [35]. However, photodynamic therapy is not effective for solid tumors, large tumors, and deep tumors. According to the principle of photodynamics, the curative effect of photodynamic therapy consists of light, photosensitizer, and oxygen. Defects from these three elements greatly limit the efficacy of photodynamic therapy in actual treatment.

Overall, photodynamic therapy, as a unique non-invasive and selective method, has been applied in the clinical treatment of diseases such as cancers. The development of nanotechnology can increase the photodynamic effectiveness and biocompatibility of photosensitizers. Besides, nanotechnology has been used to address issues such as hydrophobicity and the retention and aggregation of photosensitizers as well as improve pharmacokinetic properties and photosensitizer concentration within tumor cells. Although research using COFs for PDT is still in its early stages, these materials offer compelling advantages. For example, the absence of metal elements in their structures may improve their safety and biocompatibility; more importantly, they have excellent photosensitivity and photodynamic properties. Moreover, the PDT performance of COFs can be improved by changing the dimension, composition, and structure of COFs.

2.3 COFs used for photothermal therapy (PTT)

PTT, as an effective cancer treatment strategy, has attracted the attention of researchers. PTT absorbs NIR excitation energy through photothermal agents (PTAs). It is subsequently dissipated in the form of heat, which increases the temperature around the cancer cells and kills cancer cells [36, 37]. PTT has also become one of the most effective ways to treat tumors. Heteropoly blue (HPB) is an ideal PTA with good photothermal conversion efficiency. Wang et al. loaded HPB into COFs in situ by a one-pot method (**Figure 3a**). The resulting HPB@COF platform exhibits ideal biocompatibility, pH-responsive release properties, and high tumor-suppressive efficiency for PTT, which inhibit tumor growth. A safer and more effective COF-based nano-delivery platform is fabricated for future pH-responsive photothermal therapy [38].

Photothermal therapy ablates tumors by hyperthermia (>50°C) under laser irradiation. However, hyperthermia inevitably damages surrounding healthy tissues, which causes additional damage. Therefore, effective cancer treatment by mild photothermal therapy at low temperatures is vital. Sun et al. designed nano-agents

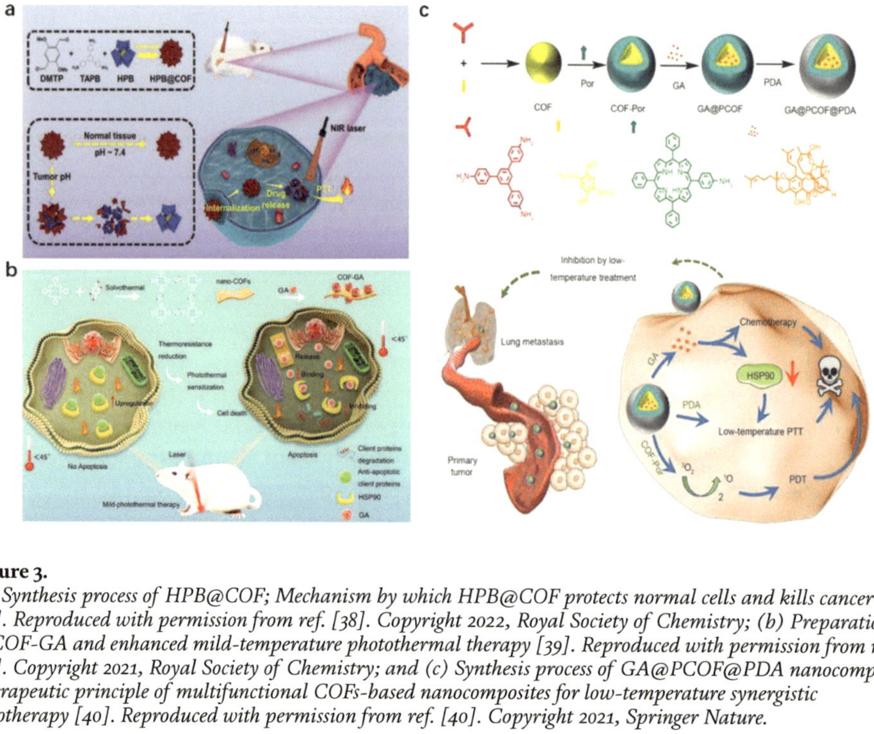

Figure 3.
(a) Synthesis process of HPB@COF; Mechanism by which HPB@COF protects normal cells and kills cancer cells [38]. Reproduced with permission from ref. [38]. Copyright 2022, Royal Society of Chemistry; (b) Preparation of COF-GA and enhanced mild-temperature photothermal therapy [39]. Reproduced with permission from ref. [39]. Copyright 2021, Royal Society of Chemistry; and (c) Synthesis process of GA@PCOF@PDA nanocomposite; Therapeutic principle of multifunctional COFs-based nanocomposites for low-temperature synergistic oncotherapy [40]. Reproduced with permission from ref. [40]. Copyright 2021, Springer Nature.

(COF-GA) to inhibit heat shock protein (HSP90), which enhances photothermal hypothermia for cancer therapy (**Figure 3b**). Nanoscale COFs can raise the temperature of tumor tissues under laser irradiation, which converts the laser light into heat to kill cancer cells. Gambogic acid (GA), as an inhibitor of HSP90, is used to overcome the thermotolerance of tumors and achieve efficient mild-temperature photothermal therapy. With the increased laser-irradiation time, COF-GA increases the temperatures at tumor sites. *In vivo* experiments demonstrate that tumor growth is significantly inhibited after COF-GA treatment. Mild photothermal therapy exhibits excellent antitumor efficacy at relatively low temperatures and minimizes nonspecific thermal damage to normal tissues [39].

Feng et al. synthesized GA@PCOF@PDA nanocomposites using stepwise bonding defect functionalization (BDF) and guests (**Figure 3c**) for encapsulation and surface modification processes of cryogenic cancer therapy. The resulting GA@P COF@ PDA reverses the thermal resistance of tumor cells by inhibiting the expression of HSP90. It is the first example of using enhanced light therapy to suppress primary and metastatic tumors at low temperatures. The approach can be further applied to other therapies involving hypothermia and broaden the development of NCOF-based multifunctional nanomedicines for safe and effective clinical applications [40].

COFs have shown great potential in catalysis and biomedicine, but it is difficult to obtain monodisperse COFs with adjustable sizes. Jing *et al.* developed a series of COFs based on electron donor-acceptor (DA) under mild conditions. Synthesized COFs exhibit excellent colloidal stability and uniform spherical morphology. Sizes can be flexibly adjusted by catalyst content, and absorption spectra also vary with sizes. By changing the electron-donating ability of the monomers can corresponding COFs possess a broad absorption spectral range, which can even be extended to the second near-infrared biological window. Obtained COFs have strong photothermal activity under laser irradiation, which inhibits tumor growth [41].

Zhao *et al.* prepared two types of 2D COFs containing naphthalene diimide (NDI) as electron acceptor (A) and triphenylamine (PT-N-COF) or triphenylbenzene (PT-B-COF) as electron donor (D). In-plane donor and acceptor units are linked by imine bonds with precise spatial distribution. The charge transfer (CT) process induced by the D-A interaction in the 2D plane results in a pronounced near-infrared absorption property. Unique structural modifications in the COF framework lead to huge differences in photophysical properties and photothermal conversion properties. Compared with PT-B-COF, PT-N-COF containing triphenylamine as donor has stronger D-A interaction and CT effect, which exhibits distinct red-shifted absorption in the NIR region. The photothermal conversion efficiency reaches 66.4%, which is in sharp contrast to 31.2% of PT-B-COF. EPR spectra confirm unpaired electrons, consistent with CT interactions in the ground state. DFT molecular-orbital simulations reveal photophysical properties and the CT process [42].

Liu *et al.* developed an efficient strategy to synthesize nanoscale DA-structured COFs with tunable sizes, long-term water dispersibility, and special selectivity recently. The designed DA structure provides channels for efficient charge-carrier transport, which endows COFs with excellent photothermal properties and greatly enhances starvation treatment efficiency. The large surface area and porosity in the cross-linked crystals allow for a high loading capacity of glucose oxidases (GOx). Surface modification with biomolecules endows it with colloidal water stability and targeting selectivity. All these good properties guarantee better tumor ablation *in vitro* and *in vivo* than monotherapy, with excellent anti-migratory effects. Detailed apoptosis studies show that combination therapy is effective in generating reactive oxygen species. The endogenous mitochondrial apoptotic pathway and Caspase-3 activation process induce apoptosis, which ultimately makes cancer cells more sensitive to insufficient energy supply and high temperatures [43].

COFs are considered to be good carriers with good biocompatibility for photothermal therapy. They are widely used due to their excellent properties such as tunability, high thermal stability, and porous structure. The effects of COFs on the photothermal system can be divided into two types. (1) COFs are taken as the carriers of photothermal material, and the effect of photothermal therapy is achieved through complexes. (2) COFs, as multifunctional carriers, can simultaneously load antitumor drugs, photosensitizers, and photothermal agents to achieve the synergistic effect of photothermal therapy and other therapies. Another type of COFs has a π-conjugated structure through structural design, which can significantly improve photothermal absorption and photostability. They are carriers and photothermal materials and have efficient reactive oxygen species (ROS) generation and photothermal-conversion ability under near-infrared light irradiation.

2.4 COFs used for immunotherapy

COFs have a larger surface area and volume, and COFs can take up a large number of molecules and deliver them to the body. Unlike amorphous polymer particles, COFs have a crystalline structure, resulting in a larger surface area to deliver molecules. Besides, the reversible covalent bonds found in COFs can be degraded in the human body to release the encapsulated molecules.

COFs can be used for protein delivery in addition to small molecules. One challenge of protein delivery is how to maintain protein activity and prevent protein denaturation during loading large amounts of protein. Importantly, the loading of COFs can be controlled by controlling pore sizes. Also, the chemical properties of COFs can be modified. Specific proteins are selectively absorbed based on protein sizes, which allows greater loading and prevents protein denaturation. The possibility of local release of proteins from COFs is important for treating autoimmune diseases and cancers.

As a form of non-inflammatory programmed cell death (PCD), the efficacy of apoptosis is often limited by apoptosis resistance in cancer cells, resulting in suboptimal therapeutic effects. Pyroptosis and ferroptosis are immunogenic PCD in contrast to apoptosis. It is a powerful anti-cancer strategy due to its favorable ability to elicit antitumor immune responses by releasing sufficient risk-associated molecular patterns (DAMPs). Zhang et al. reported a series of multi-enzyme-mimicking COFs, COF-909-Cu, COF-909-Fe, and COF-909-Ni (**Figure 4a**) as pyroptotic inducers for remodeling the tumor microenvironment, which facilitates cancer immunotherapy. Mechanistic studies suggest that these COFs can act as the steady-state destroyers of hydrogen peroxide (H_2O_2) to increase intracellular H_2O_2 levels. They exhibit excellent superoxide dismutase (SOD) mimetic activity and convert superoxide radical (O_2•-) into H_2O_2 to alleviate H_2O_2 scavenging. They also mimic the depleted glutathione (GSH) of glutathione peroxidase (GPx). The excellent photothermal therapy properties of these COFs can accelerate the Fenton-like ionization process, which enhances their chemokinetic therapy. COF-909-Cu, one of these members, strongly induces gasdermin E (GSDME)-dependent pyroptosis and remodels the tumor microenvironment. Durable anti-tumor immunity is triggered, which increases the response rate of αPD-1 checkpoint blockade and inhibits tumor metastasis and recurrence [44].

Immune checkpoint blockade therapy is revolutionizing the traditional treatment paradigm for tumors but remains ineffective for most patients. PDT has been

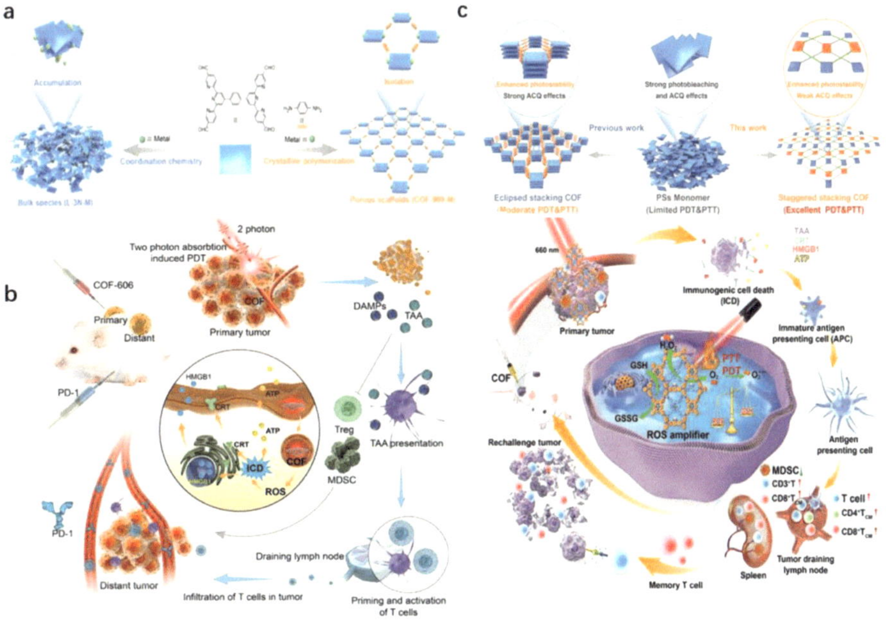

Figure 4.
(a) Engineering multienzyme-mimicking COFs as apoptosis inducers to enhance antitumor immunity [44]. Reproduced with permission from ref. [44]. Copyright 2022, Wiley-VCH; (b) Representation of COF-606-mediated 2PA-induced PDT to enhance immune checkpoint blockade therapy. COF 606-induced PDT with strong two-photon absorption, the biological process triggering immunogenic cell death (ICD), and the underlying mechanism for the synergy of PDT with PD-1 blockade therapy. TAA, the tumor-associated antigen; DAMP, the molecular pattern associated with risk [45]. Reproduced with permission from ref. [45]. Copyright 2021, Wiley-VCH; and (c) COF-618-Cu for enhancing antitumor immunity: Different structures between overlapped-stacked COFs and staggered-stacking COFs; Biological processes and underlying mechanisms of COF-618-Cu-mediated PDT and PTT for triggering cancer immunotherapy [46]. Reproduced with permission from ref. [46]. Copyright 2022, Wiley-VCH.

shown to induce cancer cell death and trigger an immune response and may represent a potential strategy for synergy with immune checkpoint blockade therapy. Yang et al. reported COF-606, a zigzag-filled COF (**Figure 4b**). Its excellent two-photon absorption (2PA) properties and photostability largely avoid aggregation-induced quenching. Therefore, it provides high ROS generation efficiency and can be used as a 2PA photosensitizer for PDT in deep tumor tissues. COF-606-induced PDT is demonstrated for the first time to be effective in inducing immunogenic cell death, eliciting an immune response, and normalizing an immunosuppressive state. 2PA-induced PDT using COF can combine with the immune checkpoint blockade therapy of programmed cell death protein 1. This combined therapeutic strategy leads to strong ectopic tumor suppression and durable immune memory effects, which is promising for cancer therapy [45].

Phototherapy-induced cancer immune responses are severely limited by the inherent photobleaching and aggregation-caused quenching (ACQ) of photosensitizers as well as intrinsic-antioxidant tumor microenvironments (TMEs), e.g., hypoxia and overexpressed glutathione (GSH). Zhang et al. designed COF-618-Cu, a novel porphyrin-based staggered-stacked COF, as an amplifier of ROS to address these issues. COF-618-Cu can consume endogenous hydrogen peroxide to generate enough oxygen because of its excellent catalase-like activity, which alleviates tumor hypoxia (**Figure 4c**). Besides, overexpressed intracellular GSH is also depleted to reduce ROS scavenging due to the glutathione peroxidase-mimicking activity of COF-618-Cu. Mechanistic studies show that the unique staggered-stacked pattern between the COF-618-Cu interlayers can alleviate the photobleaching and ACQ effects that cannot be obtained with ordinary COFs. Furthermore, COF-618-Cu, coupled with its excellent photothermal properties, is beneficial to inducing robust immunogenic cell death and remodeling TME to enhance antitumor immune effects [46].

Immunotherapy has become one of the hot topics in cancer treatment with the growing understanding of how cancer interacts with the immune system. Immunotherapy mainly attacks tumor cells by stimulating the body's innate immune system [47]. Current immunotherapies mainly include cancer vaccination [48], immune checkpoint blockade therapy [49], and chimeric antigen receptor T-cell immunotherapy (CAR-T) [50, 51]. However, solid tumors are difficult to eliminate by immunotherapy alone, so it is necessary to combine with other treatments to inhibit tumor growth and metastasis.

COFs as new photosensitizers are widely used in photothermal and photodynamic therapy of tumors. However, photothermal and photodynamic therapy using COF materials can only kill superficial tumors in irradiated areas. Triggering an immune response is difficult due to the limited ability to generate ROS, and it is still a great challenge to use COF materials to inhibit tumor metastasis and recurrence. Compared with 2D COF materials, 3D COF materials can avoid the fluorescence quenching caused by the aggregation of photosensitizers and have a penetrating pore structure conducive to ROS transport. However, the use of 3D COF materials in tumor immunotherapy has not been reported due to the difficulty of 3D COF synthesis.

2.5 Application of COFs in combination therapy for tumors

COFs have been widely used in the nano-drug delivery system of chemotherapy, photothermal therapy, and photodynamic therapy due to their dense pore structure and lipophilicity. Monotherapy is not effective in treating cancers. For example, the self-quenching of photosensitizers in the hypoxic tumor microenvironment can hinder the release of ROS [52–55], resulting in low therapeutic efficiency. However, COFs can be combined with a variety of components and modified to form a nano-drug delivery system for various combination therapies, which can obtain better therapeutic effects.

Anthocyanin (IR783), typical in the combination therapy of hyperthermia and chemotherapy, has received extensive attention recently due to its good absorption of light in the near-infrared region and good biocompatibility. Chen's research group synthesized a new type of COFs (TP-Por COFs) and exfoliated the COFs into IR783-loaded COFs nanosheets (COF@IR783) by ultrasonic exfoliation in 2019. After that, cis-aconitic anhydride-modified doxorubicin (CAD) is loaded to form a photothermal drug-chemotherapy combination therapy system (**Figure 5a**). *In vitro* cell experiments show that COFs do not affect cell viability, but the system has an obvious combined killing effect on tumor cells. Later, the tumor tissues of mice treated with this system are significantly necrotic [56].

Zhang's group combined the anti-fibrotic drug pirfenidone (PFD) with COFTTA-DHTA in the combination therapy of photodynamic and chemotherapy. Amphiphilic polymer poly (lactic-co-glycolic acid)-polyethylene glycol (PLGA-PEG) is used to synthesize PFD@COFTTA-DHTA@PLGA-PEG(PCPP) (**Figure 5b**). PFD released by PCPP in the tumor area can destroy the extracellular matrix (ECM) of tumors in

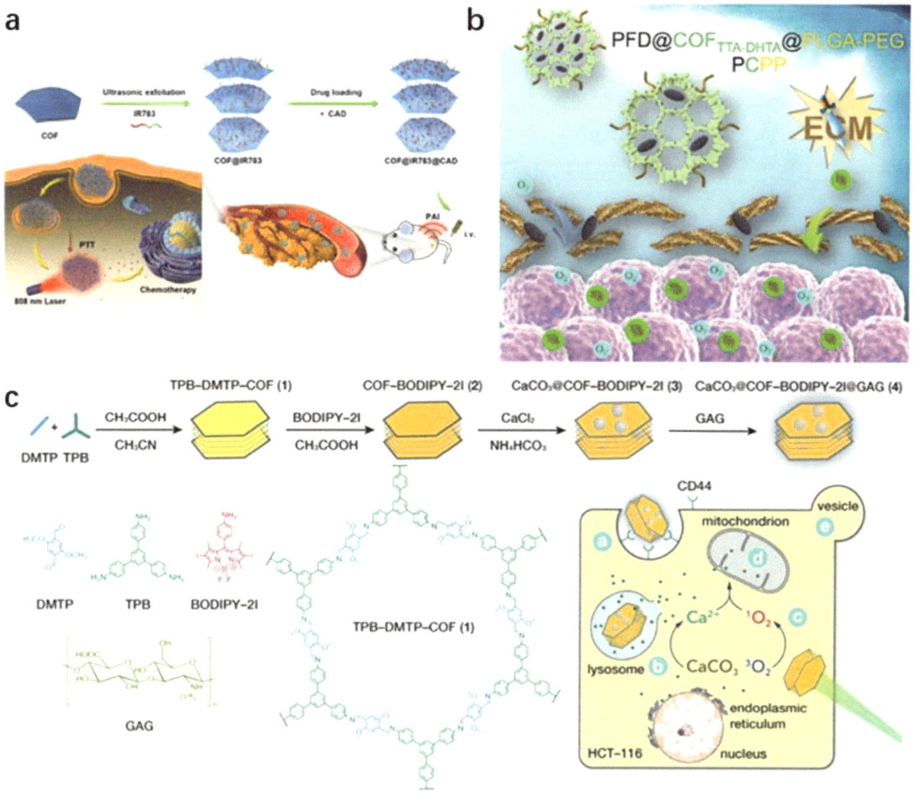

Figure 5.
(a) Applications of COFs in the combination therapy of photothermal therapy and chemotherapy and that of photodynamic therapy and chemotherapy: (A) COF@IR783@CAD in oncotherapy [56]; Reproduced with permission from ref. [56]. Copyright 2019, American Chemical Society; and (b) PFD@COFTTA-DHTA@ PLGA-PEG in oncotherapy [57]; Reproduced with permission from ref. [57]. Copyright 2020, Elsevier Ltd. (c) $CaCO_3$@COF-BODIPY-2I@GAG in oncotherapy [34]. Reproduced with permission from ref. [34]. Copyright 2020, Wiley-VCH.

this structure, which promotes the uptake of the subsequently injected photosensitizer protoporphyrin peptide coupling nano-micelles (NM-PPIX) by tumor cells and enhances the efficacy of PDT on tumors [57]. Excessive intracellular Ca^{2+} content can lead to cell death, and its concentration is tightly controlled within cells. The control is easily destroyed by ROS, so theoretically, the combined action of PDT and the addition of exogenous Ca^{2+} can promote cell death.

Inspired by this mechanism, Dong's group reported a nano-covalent organic framework (NCOF)-based nanostructure ($CaCO_3$@COF-BODIPY-2I@GAG) (**Figure 5c**). This structure uses Boron-dipyrrolemethene (BODIPY2I) as the photosensitizer and $CaCO_3$ as the material providing exogenous Ca^{2+}. ROS generated by the system under illumination can directly kill tumor cells and increase Ca^{2+} content in tumor cells to kill cells, showing a significant anti-tumor effect [34]. Trabolsi's group loaded DOX through nanoscale TAB-DFP-nCOF in the combination therapy of magnetic hyperthermia and chemotherapy. Afterward, magnetic iron oxide nanoparticles (γ-Fe_2O_3 NPs) and polylysine cationic polymer (PLL) are modified separately to obtain system γ-SD/PLL with an average particle size of 300 nm. γ-SD/PLL generates a lot of heat and releases drugs to kill cancer cells in an alternating magnetic field (AMF). The survival rate of cancer cells can be as low as 10% after 1 h of γ-SD/PLL and AMF treatment [58]. Although the development of COFs is still in its infancy, remarkable achievements have been made in chemotherapy, photothermal therapy, photodynamic therapy, and combination therapy.

Completely inhibiting tumor growth is impossible due to the complexity of the tumor microenvironment and the inherent shortcomings of monotherapy. However, combination therapy can overcome the shortcomings of monotherapy and further improve the therapeutic effect. Chen et al. synthesized COF TP-Por condensed with 5,15-bis(4-boronphenyl) porphyrin and 2,3,6,7,10,11-hexahydroxytriphenyl by a stripping method and further loaded DOX in 2019. The resulting 2D covalent organic nanosheets (CONs) can produce both PDT and PTT effects under 635-nm laser irradiation after ultrasonic stripping of TP-Por, which inhibits tumor growth [56]. Pang's group synthesized COFs based on 1,3,5-tris(4-aminophenyl) benzene (TAPB) and 1,3,5-benzenetricarbaldehyde (BTCA) by Schiff base condensation reaction for OVA and ICG loads in 2020. PDT and PTT effects occur under 650- and 808-nm laser irradiation. Combined with checkpoint blockade therapy, it shows a stronger immune effect, which can inhibit tumor metastasis and recurrence [59].

Phototherapy has recently received a lot of attention due to its simplicity, non-invasive features, and excellent therapeutic effects. The combination of PTT and PDT holds great promise in oncotherapy. A suitable photosensitizer is a prerequisite to obtaining satisfactory antitumor efficacy. Pang's group prepared highly monodisperse COF nanoparticles at room temperatures by a mild solution-phase synthesis method in 2019. The synthesized nonporphyrin-containing COF-based nanoparticles are used as novel photosensitizers for PDT and exhibit excellent photodynamic effects under 650- or 808-nm laser irradiation. Then ideal photothermal agent CuSe nanoparticles are coupled with COFs to form the bifunctional photosensitizer for phototherapy. The resulting COF-CuSe platform exhibits excellent synergistic photothermal and

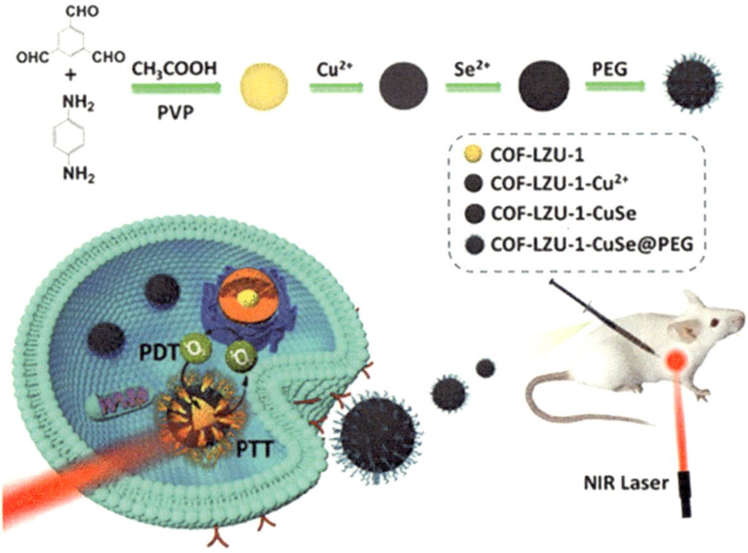

Figure 6.
(a) Synthesis of COFs-based composites and their applications in PTT and PDT [60]. Reproduced with permission from ref. [60]. Copyright 2019, American Chemical Society.

Type of COF	Characteristic synthesis	Multifunctional applications	Ref
Sor@COF-366	Simple oil-bath method	Chemotherapy /photodynamic therapy	[61]
COF-PDA-FA	Room temperature synthesis	Chemotherapy/photothermal/starvation therapy	[43]
MOF@COF	Schiff-base reaction	Microwave thermal/microwave dynamic therapy	[62]
AQ4N@THPPTK-PEG	One-pot solvothermal method	Chemotherapy/photodynamic therapy	[63]
COF-606	Co-condensation reaction	Photodynamic/immunity therapy	[45]
COF/GOx/DOX	Room temperature synthesis	Chemotherapy/photodynamic therapy	[64]
Fe_3O_4@COF-DhaTph	In situ seed growth approach	Imaging-guidedphototherapy/photodynamic therapy	[65]
COF-909-Cu, COF-909-Fe, and COF-909-Ni	Post-modification method	Photothermal/photodynamic/immunity/chemodynamic therapy	[44]
COF-Au-MnO_2	Room temperature synthesis	Photodynamic/starving-like therapy	[30]
Fe_3O_4@COF	Facile sonication associated method	Photothermal therapy/chemotherapy	[66]

Table 1.
Summarizes the COF combination therapies published in the past 2 years.

photodynamic effects. *In vitro* and *in vivo* experiments demonstrate enhanced therapeutic effects in killing cancer cells and inhibiting tumor growth. The study demonstrates the great potential of nonporphyrin-containing COFs as photosensitizer for photodynamic cancer therapy and provides a simple and effective approach. COFs are combined with other functional materials to construct COFs-based multifunctional therapeutics for cancer diagnosis and treatment (**Figure 6**) [60].

Based on the versatility of COFs, they can integrate monotherapies such as drug delivery, photodynamic therapy, and photothermal therapy to build an intelligent, multifunctional platform for targeted therapy. Combined therapy on a platform can precisely kill cancer cells based on COFs. Moreover, COFs can be metalized before or after synthesis for imaging technology, which lays the foundation for integrating diagnosis and therapy. Therefore, designing functionalized COFs-based nanomedicines for drug co-delivery and multimodal cancer therapy is a promising strategy. In addition, we also summarize the COF combination therapy published in the past 2 years, as shown in **Table 1**.

3. Conclusion

Therapeutic methods are used in cancer treatment, such as PDT, PTT, chemotherapy, immunotherapy, and targeted therapy with the understanding of cancers. PDT is a minimally invasive treatment method utilizing PSS to absorb light energy, which stimulates oxygen to generate reactive oxygen species (such as $^1O_2$, .OH, and .O^{2-}). It has good controllability, low toxicity, and low invasiveness. However, its insufficient light penetration depth and dependence on oxygen limit its development.

PTT adsorbs near-infrared-light excitation energy by PTAs. The temperature around the cancer cells rises by heat losses, which kills cancer cells. However, PTT lacks selectivity for cancer cells, so it is impossible for specific treatments. Immunotherapy works by stimulating the body's innate immune system to attack tumor cells. All of the above cancer treatments have inherent problems. That is to say, a single treatment method cannot treat tumors at all. COFs have unique advantages as emerging antitumor materials with the development of nanotechnology. Given this, the work exploited the crystalline porosity, stability, versatility, and good biocompatibility of COFs with emerging therapeutic approaches. Meanwhile, the problems existing in the preparation and treatment of COFs were improved to construct a feasible multifunctional nanosystem for cancer treatment. Despite the above examples, further research is needed on biosafety, biocompatibility, sterilization, drug loading, and controlled drug release *in vivo*. The integration of COFs with other biomaterials is expected to greatly expand the biological applications of COFs. However, the idea of applying COFs to tumor therapy is still far from clinical translation. The main challenges and opportunities are as follows:

1. There is no universal method for mass production of COFs, because low yields and poor-quality consistency may arise in mass production. The currently available methods, including assisted solvothermal methods, steric-induced chemical exfoliation methods, and intercalation-induced delamination methods, may only be matched to specific species of COFs and high-power mechanical layering has low output and is not suitable for industrial production.

2. The safety of COFs in tumor therapy still needs to be confirmed. Current researches mostly focus on materials uptake rather than metabolic pathways due to

the short development history. COFs are usually composed of non-metallic light elements and irreversible chemical bonds. These chemical bonds ensure the stability of COFs; however, they may cause the accumulation of COFs in the body, and the toxicity of long-term use is still unpredictable. Some easily degradable COFs have potential to generate aromatic compounds in the body and cause *in-vivo* toxicity. The current safety evaluation of COFs-based oncotherapy is mostly based on cytotoxicity, a comprehensive assessment is required for their hemocompatibility, histocompatibility, cytotoxicity, neurotoxicity, and genotoxicity at cellular and tissue levels.

3. Cancer is a complex, tricky disease. There is an urgent need to develop an intelligent and multifunctional therapeutic platform for diagnosis and targeted therapy to realize COFs-based precision cancer therapy. Besides drug delivery, photodynamic therapy, and photothermal therapy, COFs can be used for combined therapy, which kills most tumor cells in multiple directions and layers. Besides, COFs possess the properties of ordered porous nanomaterials and can be easily modified. For example, metallization can be used in imaging techniques such as two-photon fluorescence imaging either before or after synthesis, with the opportunity to integrate diagnoses and treatment. The pre- and post-synthesis of COFs can be modified to increase their targeting, water solubility, etc., for better oncotherapy. It is beneficial for building a smart therapy platform based on COFs.

In summary, COFs have unique applications in nanomedicine due to their easy synthesis, large pore size, tunable structure, versatility, chemical stability, and good biocompatibility. The work briefly discussed recent advances in COFs-based drug delivery and synergistic therapy. The application of COFs in biomedicine has yielded some exciting results. Their application as nano-drugs in biomedicine is greatly limited due to the difficulty of precise control of the size and structure of COFs-based nano-platforms as well as their poor dispersibility in water. Besides, the targeting of COFs-based nano-loading systems needs to be further studied. COFs require further synthesis to achieve the pore geometries and precise crystal structures required for specific biomedical applications (e.g., protein delivery) after considering emerging approaches such as bottom-up and top-down approaches.

Acknowledgements

This chapter was financially supported by the Basic and Clinical Collaborative Research Promotion Program of Anhui Medical University (No. 2021xkjT024) and the Ph.D. Start-up Fund of Anhui Medical University.

Conflict of interest

The authors declare no conflict of interest.

Author details

Guiyang Zhang
Department of Pharmacology, School of Basic Medical Sciences, Anhui Medical University, Hefei, China

*Address all correspondence to: guiyangzhang@ahmu.edu.cn

References

[1] Das S, Heasman P, Ben T, Qiu S. Porous organic materials: Strategic design and structure-function correlation. Chemical Reviews. 2017;**117**:1515-1563. DOI: 10.1021/acs.chemrev.6b00439

[2] Ahuja G, Pathak K. Porous carriers for controlled/modulated drug delivery. Indian Journal of Pharmaceutical Sciences. 2009;**71**:599-607. DOI: 10.4103/0250-474X.59540

[3] Ma H, Hu J, Ma PX. Polymer scaffolds for small-diameter vascular tissue engineering. Advanced Functional Materials. 2010;**20**:2833-2841. DOI: 10.1002/adfm.201000922

[4] Shivanand P, Sprockel OL. A controlled porosity drug delivery system. International Journal of Pharmaceutics. 1998;**167**:83-96. DOI: 10.1016/S0378-5173(98)00047-7

[5] Acharya AP, Lewis JS, Keselowsky BG. Combinatorial co-encapsulation of hydrophobic molecules in poly(lactide-co-glycolide) microparticles. Biomaterials. 2013;**34**:3422-3430. DOI: 10.1016/j.biomaterials.2013.01.032

[6] Ratay ML, Glowacki AJ, Balmert SC, Acharya AP, Polat J, Andrews LP, et al. Treg-recruiting microspheres prevent inflammation in a murine model of dry eye disease. Journal of Controlled Release. 2017;**258**:208-217. DOI: 10.1016/j.jconrel.2017.05.007

[7] Acharya AP, Carstens MR, Lewis JS, Dolgova N, Xia CQ, Clare-Salzler MJ, et al. A cell-based microarray to investigate combinatorial effects of microparticle-encapsulated adjuvants on dendritic cell activation. Journal of Materials Chemistry B. 2016;**4**:1672-1685. DOI: 10.1039/C5TB01754H

[8] Yaghi O, O'Keeffe M, Ockwig N, Chae H, Eddaoudi M, Kim J. Reticular synthesis and the design of new materials. Nature. 2003;**423**:705-714. DOI: 10.1038/nature01650

[9] Furukawa H, Cordova KE, O'Keeffe M, Yaghi OM. The chemistry and applications of metal-organic frameworks. Science. 2013;**341**:1230444. DOI: 10.1126/science.1230444

[10] Nishiyama N, Kataoka K. Current state, achievements, and future prospects of polymeric micelles as nanocarriers for drug and gene delivery. Pharmacology&Therapeutics. 2006;**112**:630-648. DOI: 10.1016/j.pharmthera.2006.05.006

[11] Sharma RI, Pereira M, Schwarzbauer JE, Moghe PV. Albumin-derived nanocarriers: Substrates for enhanced cell adhesive ligand display and cell motility. Biomaterials. 2006;**27**:3589-3598. DOI: 10.1016/j.biomaterials.2006.02.007

[12] Horcajada P, Chalati T, Serre C, Gillet B, Sebrie C, Baati T, et al. Porous metal-organic-framework nanoscale carriers as a potential platform for drug delivery and imaging. Nature Materials. 2010;**9**:172-178. DOI: 10.1038/nmat2608

[13] Cote AP, Benin AI, Ockwig NW, O'Keeffe M, Matzger AJ, Yaghi OM. Porous, crystalline, covalent organic frameworks. Science. 2005;**310**:1166-1170. DOI: 10.1126/science.1120411

[14] Zhou S, Meng T, Hu D, Zhu Y, Huang C, Song M, et al. Characteristic synthesis of a covalent organic framework and its application in multifunctional tumor therapy. ACS Applied Bio Materials. 2022;**5**:59-81. DOI: 10.1021/acsabm.1c01039

[15] Segura JL, Mancheno MJ, Zamora F. Covalent organic frameworks based on Schiff-base chemistry: Synthesis, properties and potential applications. Chemical Society Reviews. 2016;**45**:5635-5671. DOI: 10.1039/c5cs00878f

[16] Ding SY, Wang W. Covalent organic frameworks (COFs): From design to applications. Chemical Society Reviews. 2013;**42**:548-568. DOI: 10.1039/c2cs35072f

[17] Guan Q, Zhou LL, Li YA, Li WY, Wang S, Song C, et al. Nanoscale covalent organic framework for combinatorial antitumor photodynamic and photothermal therapy. ACS Nano. 2019;**13**:13304-13316. DOI: 10.1021/acsnano.9b06467

[18] Bhunia S, Deo KA, Gaharwar AK. 2D Covalent organic frameworks for biomedical applications. Advanced Functional Materials. 2020;**30**:2002046. DOI: 10.1002/adfm.202002046

[19] Tao D, Feng L, Chao Y, Liang C, Song X, Wang H, et al. Covalent organic polymers based on fluorinated porphyrin as oxygen nanoshuttles for tumor hypoxia relief and enhanced photodynamic therapy. Advanced Functional Materials. 2018;**28**:1804901. DOI: 10.1002/adfm.201804901

[20] Zhang L, Wang S, Zhou Y, Wang C, Zhang XZ, Deng H. Covalent organic frameworks as favorable constructs for photodynamic therapy. Angewandte Chemie International Edition. 2019;**58**:14213-14218. DOI: 10.1002/anie.201909020

[21] Zheng X, Wang L, Guan Y, Pei Q, Jiang J, Xie Z. Integration of metal-organic framework with a photoactive porous-organic polymer for interface enhanced phototherapy. Biomaterials. 2020;**235**:119792. DOI: 10.1016/j.biomaterials.2020.119792

[22] Halim J, Lukatskaya MR, Cook KM, Lu J, Smith CR, Naslund LA, et al. Transparent conductive two-dimensional titanium carbide epitaxial thin films. Chemical Materials. 2014;**26**:2374-2381. DOI: 10.1021/cm500641a

[23] Liu S, Liu Y, Hu C, Zhao X, Ma P, Pang M. Boosting the antitumor efficacy over a nanoscale porphyrin-based covalent organic polymer via synergistic photodynamic and photothermal therapy. Chemical Communication. 2019;**55**:6269-6272. DOI: 10.1039/c9cc02345c

[24] Bai L, Phua SZ, Lim WQ, Jana A, Luo Z, Tham HP, et al. Nanoscale covalent organic frameworks as smart carriers for drug delivery. Chemical Communication. 2016;**52**:4128-4131. DOI: 10.1039/c6cc00853d

[25] Mitra S, Sasmal HS, Kundu T, Kandambeth S, Illath K, Díaz Díaz D, et al. Targeted drug delivery in covalent organic nanosheets (cons) via sequential postsynthetic modification. Journal of the American Chemical Society. 2017;**139**:4513-4520. DOI: 10.1021/jacs.7b00925

[26] Zhang G, Li X, Liao Q, Liu Y, Xi K, Huang W, et al. Water-dispersible PEG-curcumin/amine-functionalized covalent organic framework nanocomposites as smart carriers for in vivo drug delivery. Nature Communications. 2018;**9**:2785. DOI: 10.1038/s41467-018-04910-5

[27] Zhang G, Jiang B, Wu C, Liu Y, He Y, Huang X, et al. Thin platelet-like COF nanocomposites for blood brain barrier transport and inhibition of brain metastasis from renal cancer. Journal of Materials Chemistry B. 2020;**8**:4475-4488. DOI: 10.1039/d0tb00724b

[28] Tong X, Gan S, Wu J, Hu Y, Yuan A. A nano-photosensitizer based on covalent organic framework nanosheets with high loading and therapeutic efficacy. Nanoscale. 2020;**12**:7376-7382. DOI: 10.1039/c9nr10787h

[29] Liu S, Hu C, Liu Y, Zhao X, Pang M, Lin J. One-pot synthesis of DOX@Covalent organic framework with enhanced chemotherapeutic efficacy. Chemistry. 2019;**25**:4315-4319. DOI: 10.1002/chem.201806242

[30] Cai L, Hu C, Liu S, Zhou Y, Pang M, Lin J. A covalent organic framework-based multifunctional therapeutic platform for enhanced photodynamic therapy via catalytic cascade reactions. Science China Materials. 2020;**64**:488-497. DOI: 10.1007/s40843-020-1428-0

[31] Zhang Y, Zhang L, Wang Z, Wang F, Kang L, Cao F, et al. Renal-clearable ultrasmall covalent organic framework nanodots as photodynamic agents for effective cancer therapy. Biomaterials. 2019;**223**:119462. DOI: 10.1016/j.biomaterials.2019.119462

[32] Gao P, Wang M, Chen Y, Pan W, Zhou P, Wan X, et al. A COF-based nanoplatform for highly efficient cancer diagnosis, photodynamic therapy and prognosis. Chemical Science. 2020;**11**:6882-6888. DOI: 10.1039/d0sc00847h

[33] Wang P, Zhou F, Guan K, Wang Y, Fu X, Yang Y, et al. In vivo therapeutic response monitoring by a self-reporting upconverting covalent organic framework nanoplatform. Chemical Science. 2019;**11**:1299-1306. DOI: 10.1039/c9sc04875h

[34] Guan Q, Zhou LL, Lv FH, Li WY, Li YA, Dong YB. A glycosylated covalent organic framework equipped with BODIPY and CaCO3 for synergistic tumor therapy. Angewandte Chemie International Edition. 2020;**59**:18042-18047. DOI: 10.1002/anie.202008055

[35] Huang Z. A review of progress in clinical photodynamic therapy. Technology in Cancer Research & Treatment. 2005;**4**:283-293. DOI: 10.1177/153303460500400308

[36] Chen Y, Li L, Chen W, Chen H, Yin J. Near-infrared small molecular fluorescent dyes for photothermal therapy. Chinese Chemical Letters. 2019;**30**:1353-1360. DOI: 10.1016/j.cclet.2019.02.003

[37] Wei W, Zhang X, Zhang S, Wei G, Su Z. Biomedical and bioactive engineered nanomaterials for targeted tumor photothermal therapy: A review. Materials Science & Engineering C. 2019;**104**:109891. DOI: 10.1016/j.msec.2019.109891

[38] Wang W, Song Y, Chen J, Yang Y, Wang J, Song Y, et al. Polyoxometalate-covalent organic framework hybrid materials for pH-responsive photothermal tumor therapy. Journal of Materials Chemistry B. 2022;**10**:1128-1135. DOI: 10.1039/d1tb02255e

[39] Sun Q, Tang K, Song L, Li Y, Pan W, Li N, et al. Covalent organic framework based nanoagent for enhanced mild-temperature photothermal therapy. Biomaterials Science. 2021;**9**:7977-7983. DOI: 10.1039/d1bm01245b

[40] Feng J, Ren W-X, Kong F, Zhang C, Dong Y-B. Nanoscale covalent organic framework for the low-temperature treatment of tumor growth and lung metastasis. Science China Materials. 2021;**65**:1122-1133. DOI: 10.1007/s40843-021-1805-x

[41] Xia R, Zheng X, Li C, Yuan X, Wang J, Xie Z, et al. Nanoscale covalent

organic frameworks with donor-acceptor structure for enhanced photothermal ablation of tumors. ACS Nano. 2021;**15**:7638-7648. DOI: 10.1021/acsnano.1c01194

[42] Zhang Y, Wu G, Liu H, Tian R, Li Y, Wang D, et al. Donor–acceptor based two-dimensional covalent organic frameworks for near-infrared photothermal conversion. Materials Chemistry Frontiers. 2021;**5**:6575-6581. DOI: 10.1039/d1qm00462j

[43] Song S, Wang D, Zhao K, Wu Y, Zhang P, Liu J, et al. Donor-acceptor structured photothermal COFs for enhanced starvation therapy. Chemical Engineering Journal. 2022;**442**:135963. DOI: 10.1016/j.cej.2022.135963

[44] Zhang L, Yang QC, Wang S, Xiao Y, Wan SC, Deng H, et al. Engineering multienzyme-mimicking covalent organic frameworks as pyroptosis inducers for boosting antitumor immunity. Advanced Materials. 2022;**34**:e2108174. DOI: 10.1002/adma.202108174

[45] Yang LL, Zhang L, Wan SC, Wang S, Wu ZZ, Yang QC, et al. Two-photon absorption induced cancer immunotherapy using covalent organic frameworks. Advanced Functional Materials. 2021;**31**:2103056. DOI: 10.1002/adfm.202103056

[46] Zhang L, Xiao Y, Yang QC, Yang LL, Wan SC, Wang S, et al. Staggered stacking covalent organic frameworks for boosting cancer immunotherapy. Advanced Functional Materials. 2022;**32**:2201542. DOI: 10.1002/adfm.202201542

[47] Carreno BMMV, Becker-Hapak M, Kaabinejadian S, Hundal J, Petti AA, Ly A, et al. A dendritic cell vaccine increases the breadth and diversity of melanoma neoantigen-specific T cells. Science. 2015;**348**:803-808. DOI: 10.1126/science.aaa3828

[48] Palucka K, Banchereau J. Cancer immunotherapy via dendritic cells. Nature Review Cancer. 2012;**12**:265-277. DOI: 10.1038/nrc3258

[49] Pardoll DM. The blockade of immune checkpoints in cancer immunotherapy. Nature Review Cancer. 2012;**12**:252-264. DOI: 10.1038/nrc3239

[50] Restifo NP, Dudley ME, Rosenberg SA. Adoptive immunotherapy for cancer: Harnessing the T cell response. Nature Reviews Immunology. 2012;**12**:269-281. DOI: 10.1038/nri3191

[51] Chen Q, Chen M, Liu Z. Local biomaterials-assisted cancer immunotherapy to trigger systemic antitumor responses. Chemical Society Reviews. 2019;**48**:5506-5526. DOI: 10.1039/c9cs00271e

[52] Li X, Kwon N, Guo T, Liu Z, Yoon J. Innovative strategies for hypoxic-tumor photodynamic therapy. Angewandte Chemie International Edition. 2018;**57**:11522-11531. DOI: 10.1002/anie.201805138

[53] Chen Z, Liu L, Liang R, Luo Z, He H, Wu Z, et al. Bioinspired hybrid protein oxygen nanocarrier amplified photodynamic therapy for eliciting anti-tumor immunity and abscopal effect. ACS Nano. 2018;**12**:8633-8645. DOI: 10.1021/acsnano.8b04371

[54] Reisch A, Klymchenko AS. Fluorescent polymer nanoparticles based on dyes: Seeking brighter tools for bioimaging. Small. 2016;**12**:1968-1992. DOI: 10.1002/smll.201503396

[55] Yang G, Xu L, Chao Y, Xu J, Sun X, Wu Y, et al. Hollow MnO2 as a

tumor-microenvironment-responsive biodegradable nano-platform for combination therapy favoring antitumor immune responses. Nature Communications. 2017;**8**:902. DOI: 10.1038/s41467-017-01050-0

[56] Wang K, Zhang Z, Lin L, Hao K, Chen J, Tian H, et al. Cyanine-assisted exfoliation of covalent organic frameworks in nanocomposites for highly efficient chemo-photothermal tumor therapy. ACS Applied Materials & Interfaces. 2019;**11**:39503-39512. DOI: 10.1021/acsami.9b13544

[57] Wang SB, Chen ZX, Gao F, Zhang C, Zou MZ, Ye JJ, et al. Remodeling extracellular matrix based on functional covalent organic framework to enhance tumor photodynamic therapy. Biomaterials. 2020;**234**:119772. DOI: 10.1016/j.biomaterials.2020.119772

[58] Benyettou F, Das G, Nair AR, Prakasam T, Shinde DB, Sharma SK, et al. Covalent organic framework embedded with magnetic nanoparticles for MRI and chemo-thermotherapy. Journal of the American Chemical Society. 2020;**142**:18782-18794. DOI: 10.1021/jacs.0c05381

[59] Zhou Y, Liu S, Hu C, Cai L, Pang M. A covalent organic framework as a nanocarrier for synergistic phototherapy and immunotherapy. Journal of Materials Chemistry B. 2020;**8**:5451-5459. DOI: 10.1039/d0tb00679c

[60] Hu C, Zhang Z, Liu S, Liu X, Pang M. Monodispersed CuSe sensitized covalent organic framework photosensitizer with an enhanced photodynamic and photothermal effect for cancer therapy. ACS Applied Materials & Interfaces. 2019;**11**:23072-23082. DOI: 10.1021/acsami.9b08394

[61] Hu J, Hu J, Wu W, Qin Y, Fu J, Liu C, et al. Bimodal treatment of hepatocellular carcinoma by targeted minimally interventional photodynamic/chemotherapy using glyco-covalent-organic frameworks-guided porphyrin/sorafenib. Acta Biomaterialia. 2022;**148**:206-217. DOI: 10.1016/j.actbio.2022.06.012

[62] Li S, Chen Z, Tan L, Wu Q, Ren X, Fu C, et al. MOF@COF nanocapsule for the enhanced microwave thermal-dynamic therapy and anti-angiogenesis of colorectal cancer. Biomaterials. 2022;**283**:121472. DOI: 10.1016/j.biomaterials.2022.121472

[63] He H, Du L, Xue H, Wu J, Shuai X. Programmable therapeutic nanoscale covalent organic framework for photodynamic therapy and hypoxia-activated cascade chemotherapy. Acta Biomaterialia. 2022;**147**:1-440. DOI: Acta Biomaterialia

[64] Xu J, Xu Y, Sun L, Lu B, Yan X, Wang Z, et al. Glucose oxidase loaded Cu(2+) based metal-organic framework for glutathione depletion/reactive oxygen species elevation enhanced chemotherapy. Biomedicine & Pharmacotherapy. 2021;**141**:111606. DOI: 10.1016/j.biopha.2021.111606

[65] Feng J, Ren WX, Gao JL, Li F, Kong F, Yao BJ, et al. Core-shell-structured covalent-organic framework as a nanoagent for single-laser-induced phototherapy. ACS Applied Materials & Interfaces. 2021;**13**:17243-17254. DOI: 10.1021/acsami.1c01125

[66] Zhao K, Gong P, Huang J, Huang Y, Wang D, Peng J, et al. Fluorescence turn-off magnetic COF composite as a novel nanocarrier for drug loading and targeted delivery. Microporous and Mesoporous Materials. 2021;**311**:110713. DOI: 10.1016/j.micromeso.2020.110713